HIGH ENERGY PHYSICS AT THE MILLENNIUM: MRST '99

HIGH ENERGY PHYSICS AT THE MILLENNIUM: MRST '99

"The Sundarfest"

Ottawa, Ontario   May 1999

EDITORS
Pat Kalyniak
Stephen Godfrey
Basim Kamal
Carleton University, Ottawa

American Institute of Physics

AIP CONFERENCE
PROCEEDINGS 488

Melville, New York

Editors:

Pat Kalyniak
Stephen Godfrey
Basim Kamal

Physics Department
Carleton University
1125 Colonel By Drive
Ottawa, Ontario K1S 5B6
CANADA

E-mail: kalyniak@physics.carleton.ca
godfrey@physics.carleton.ca
bkamal@physics.carleton.ca

Authorization to photocopy items for internal or personal use, beyond the free copying permitted under the 1978 U.S. Copyright Law (see statement below), is granted by the American Institute of Physics for users registered with the Copyright Clearance Center (CCC) Transactional Reporting Service, provided that the base fee of $15.00 per copy is paid directly to CCC, 222 Rosewood Drive, Danvers, MA 01923. For those organizations that have been granted a photocopy license by CCC, a separate system of payment has been arranged. The fee code for users of the Transactional Reporting Service is: 1-56396-902-5/99/$15.00.

© 1999 American Institute of Physics

Individual readers of this volume and nonprofit libraries, acting for them, are permitted to make fair use of the material in it, such as copying an article for use in teaching or research. Permission is granted to quote from this volume in scientific work with the customary acknowledgment of the source. To reprint a figure, table, or other excerpt requires the consent of one of the original authors and notification to AIP. Republication or systematic or multiple reproduction of any material in this volume is permitted only under license from AIP. Address inquiries to Office of Rights and Permissions, Suite 1NO1, 2 Huntington Quadrangle, Melville, N.Y. 11747-4502; phone: 516-576-2268; fax: 516-576-2499; e-mail: rights@aip.org.

L.C. Catalog Card No. 99-066537
ISBN 1-56396-902-5
ISSN 0094-243X
DOE CONF- 990517

Printed in the United States of America

Contents

Preface vii
Sponsors ix
Photograph xi

Searching for a W' at the Next-Linear-Collider Using Single Photons 1
    S. Godfrey, P. Kalyniak, B. Kamal, and A. Leike

Closing the Low-mass Axigluon Window 9
    M. A. Doncheski

Lowest-Lying Scalar Mesons and a Possible Probe of Their Quark Substructure 17
    A. H. Fariborz

The Theta Term in QCD Sum Rules and Electric Dipole Moments of Hadrons 25
    M. Pospelov and A. Ritz

The Supersymmetric Spectrum of Gauge-Mediated Supersymmetry Breaking of $SO(10)$ 33
    M. Frank, H. Hamidian, and K. Puolamäki

Ultrastrong Coupling in Supersymmetric Gauge Theories 40
    A. Buchel

Lepton–Chargino Mixing and R-Parity Violating SUSY 49
    M. Bisset, O. C. W. Kong, C. Macesanu, and L. H. Orr

Flavour Changing Neutral Currents in Supersymmetric Models with Large $\tan\beta$ 56
    C. Hamzaoui, M. Pospelov, A. Raymond, and M. Toharia

Inflation from Extra Dimensions 64
    J. M. Cline

Large-N Yang-Mills Theory as Classical Mechanics 72
    C.-W. H. Lee and S. G. Rajeev

Gauge Field Correlators in 4-D Simplicial Quantum Gravity 80
    E. B. Gregory and S. M. Catterall

The N-Cosine Model—Algebraic Structures and Integrable Points on the Marginal Manifold 86
    B. Gerganov

Coherent States in High-Energy Physics 95
    C. S. Lam

The Ising Model on a Fluctuating Disk 103
    S. V. McGuire and S. M. Catterall

Coherent Conversion of Neutrino Flavour by Collisions with Relic Neutrino Gas 110
    I. S. Batkin and M. K. Sundaresan

Local and Global Duality and the Determination of $\alpha(M_Z)$ 124
    S. Groote

Looking for New Physics in $B_d^0 - \bar{B}_d^0$ Mixing 132
    D. London

T-odd Triple Product Asymmetries in Beauty Decays 140
    W. Bensalem
Study of $B \to D^{(*)+}D^{(*)-}K_s$ Decays and the Extraction of β 146
    T. E. Browder, A. Datta, P. J. O'Donnell, and S. Pakvasa
Infrared Effects in the Decay $B \to X_s \ell^+ \ell^-$ 154
    C. W. Bauer
Predictions for the Semi-Leptonic and Non-Leptonic Decays of the Λ_b and B Meson 161
    A. Datta, H. J. Lipkin, and P. J. O'Donnell
Multiplicities of Gluon and Quark Jets in pQCD 169
    I. M. Dremin
Continuum Background Suppression Using Various Selectors 174
    M. Milek and P. M. Patel
New Physics in Top Polarization at the Tevatron 182
    T. Torma
Gluons in a Color-Neutral Nucleus 190
    G. Mahlon
Quark- and Gluon-Condensate Contributions to Penguin Four-Fermi Operators 198
    M. R. Ahmady and V. Elias
An Investigation of Nuclear Collisions with a Momentum-dependent Lattice Hamiltonian Model 206
    D. Persram and C. Gale
Lepton Pair Emission Rates from a Hot Hadron Gas 214
    I. Kvasnikova and C. Gale
The Role of Baryons in the Production of Dileptons during Relativistic Heavy Ion Collisions 222
    A. K. Dutt-Mazumder, C. Gale, C. M. Ko, and V. Koch

List of Participants 231
Schedule 233
Author Index 237

Preface

The 21st Annual Montreal-Rochester-Syracuse-Toronto (MRST) Conference on High Energy Physics was held May 10-12, 1999 at Carleton University in Ottawa. This was the first time the Conference was held outside its "initial" cities and is indicative of the broad participation in MRST from east-central Canada and the northeastern United States. The venue allowed us the opportunity to celebrate the career in High Energy Physics of our Carleton colleague, Prof. M.K. Sundaresan. Hence, the conference was subtitled 'THE SUNDARFEST'.

M.K. Sundaresan has made wide ranging contributions throughout his lifetime of work in theoretical physics. Sundar's greatest strength is the breadth of his knowledge and this is reflected in his achievements. His work ranges over plasma, atomic, nuclear, and elementary particle physics.

Sundar made the first applications, to meson-nucleon scattering, of Tamm-Dancoff theory in 1954. This methodology has enjoyed a revival so Sundar's current research program includes the use of the Light Front Tamm-Dancoff formulation in Quantum Chromodynamics to deal with hadronic bound states. He is also recognized for his work in muonic atoms, wherein he cleared up a major discrepancy between the existing theoretical predictions and experimental measurements of the transition energies. Sundar made early progress in the phenomenology of the Higgs boson with collaborators, calculating the decay rate into two photons. This mode is a leading contender for the discovery of a Higgs boson. He has lately considered environmental effects on beta decay, including corrections to tritium decay. Also, in a solar plasma, interactions with plasmons significantly change the relative probabilities of different nuclear processes. Both have implications for neutrino physics and the solar model.

At the MRST banquet, Sundar's accomplishments were warmly and eloquently praised by Faqir Khanna and C.S. Lam.

The conference lived up to its usual reputation in a couple of ways. First, it is characterized by a very broad range of topics including heavy quark physics, neutrinos, collider physics, supersymmetry and other extensions of the Standard Model, heavy ion collisions, gravity, and field theoretic methods. Secondly, MRST always features many contributions from postdocs and graduate students.

Thanks to Eva Lacelle for her help in making arrangements for the smooth running of the conference. Finally, we would like to thank all the participants for coming to MRST at Carleton.

Pat Kalyniak
Stephen Godfrey
Basim Kamal

August 1999
Ottawa, Canada

Sponsors

We gratefully acknowledge the generous support of our conference sponsors.

- Carleton University Faculty of Science
- Carleton University Office of the Vice-President Research and External
- The Institute of Particle Physics
- The Canadian Association of Physics Particle Physics Division
- The Ottawa-Carleton Institute for Physics
- Cartier Place and Towers Suite Hotels

M. K. Sundaresan

Searching for a W' at the Next-Linear-Collider using Single Photons*

Stephen Godfrey, Pat Kalyniak and Basim Kamal

*Ottawa-Carleton Institute for Physics
Department of Physics, Carleton University, Ottawa, ON K1S 5B6, Canada*

Arnd Leike

LMU, Sektion Physik, Theresienstr. 37, D-80333 München, Germany

Abstract. We examine the sensitivity of the process $e^+e^- \to \nu\bar{\nu} + \gamma$ to additional W-like bosons which arise in various models. The process is found to be sensitive to W' masses up to several TeV.

INTRODUCTION

There have been many studies of processes sensitive to additional Z-like bosons (Z's) but comparatively few studies pertinent to W's (see for instance [1,2]), especially at e^+e^- colliders. Why? Firstly, there are fewer models which predict W's. Secondly, at LEP II energies, the W' signal is rather weak. Hence direct searches have been limited to hadron colliders and the W'-quark couplings are poorly constrained in some models.

In this contribution we present preliminary results of an investigation of the sensitivity of the process $e^+e^- \to \nu\bar{\nu} + \gamma$ to W' bosons in various models. Our results are obtained by measuring the deviation from the standard model expectation. Interesting discovery limits are obtained for center-of-mass e^+e^- energy of 500 GeV or higher – *Next-Linear-Collider* (NLC) energies.

Direct searches have been performed and indirect bounds have been obtained for W's in a few models, the details of which are given later. Bounds for the Left-Right Symmetric Model (LRM) and Sequential Standard Model (SSM) can be found in [3]. They are obtained from the non-observation of direct production of W's at the Tevatron and from indirect μ-decay constraints. For the LRM (with equal left- and right-handed couplings) CDF obtains $M_{W'} \gtrsim 650$ GeV and for the SSM, D0 finds

---

*) Supported in part by the Natural Sciences and Engineering Research Council of Canada.

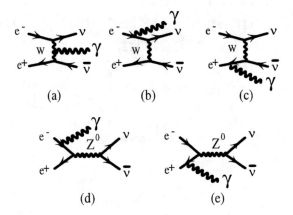

FIGURE 1. The Feynman diagrams contributing to the SM process.

$M_{W'} \gtrsim 720$ GeV. From μ-decay, the LRM W' is constrained to $M_{W'} \gtrsim 550$ GeV [4]. A naive leading order analysis for the SSM yields a bound of between 900 GeV and 1 TeV. One expects a somewhat higher bound than for the LRM since, in μ-decay, there will be a W-W' interference term. The major limitation in this method is the uncertainty in the W mass.

The LHC will have a discovery reach in the TeV range [5]. The search is analogous to that done at the Tevatron, except for the higher energy and luminosity. On the down side, one has pp instead of $p\bar{p}$, which means no valence-valence contribution in the large Feynman-x region. For the LRM, the magnitude of the effect will depend also on the W'-quark couplings which are unknown. Therefore it is hard to make predictions a-priori concerning discovery limits. The NLC search nicely sidesteps the above problem as no W'-quark couplings enter. Other LHC disadvantages include a lack of initial state quark polarizability, parton distribution dependence and large QCD corrections. The latter problems will affect the ability to pin down the W' couplings. Hence, the complimentary nature of the cleaner NLC measurement is obvious despite the LHC's high energy reach.

BASIC PROCESS

The basic process under consideration is:

$$e^-(p_1) + e^+(p_2) \rightarrow \gamma(p_3) + [\nu(p_4) + \bar{\nu}(p_5)], \tag{1}$$

where the square brackets indicate that since the neutrinos are not observed, we effectively only have single photon production. The diagrams representing the leading Standard Model (SM) contribution are shown in Fig. 1. The $W'(Z')$ contributions are obtained by replacing $W \rightarrow W'$, $Z \rightarrow Z'$ in the SM diagrams. Then one must include all interferences between SM and beyond-SM diagrams in the squared amplitude.

The resulting squared amplitude is quite short, including spin dependence which comes out automatically when expressing the result in terms of the left- and right-handed couplings. The result is quite general and includes an arbitrary number of W's and Z's.

MODELS

We have considered three models having W's which contribute to our process and are briefly described below.

Sequential Standard Model: This is the simplest W'-containing extension of the SM, although not well motivated by theory. One has an extra W' which is heavier than the SM one, but which has identical couplings.

Left-Right Symmetric Model: In this model [6], the symmetry $SU(3)_c \times SU(2)_L \times SU(2)_R \times U(1)_{B-L}$ is obeyed, giving rise to a W' and a Z'. The W' is purely right-handed; we do not consider mixing between the SM and beyond-SM bosons. The pure SM couplings remain unchanged and we take the new right-handed neutrinos to be massless. In principle they could be very heavy as well, but this would lead to decoupling of the W' from our process and we would be effectively left with a Z' model, which is not the principal interest of this study.

Two parameters arise; ρ and κ. For symmetry breaking via Higgs doublets (triplets) $\rho = 1$ (2). κ is defined by $\kappa = g_R/g_L$ and thus measures the relative strength of the $W'l\nu_l$ and $Wl\nu_l$ couplings. It lies in the range [2,7]

$$0.55 \lesssim \kappa \lesssim 2 . \tag{2}$$

More specifically, we have the coupling

$$W'l\nu = i\frac{g\kappa}{\sqrt{2}}\gamma^\mu \frac{1+\gamma_5}{2}, \tag{3}$$

suggesting that larger values of κ will lead to larger deviations from the SM. In addition, we have the relation

$$M_{Z'}^2 = \frac{\rho\kappa^2}{\kappa^2 - \tan^2\theta_W} M_{W'}^2, \tag{4}$$

so that $\rho = 1$ leads to a lighter Z' mass for fixed $M_{W'}$, which should yield a bigger effect versus $\rho = 2$.

Un-Unified Model (UUM): The UUM [8] obeys the symmetry $SU(2)_q \times SU(2)_l \times U(1)_Y$, again leading to a W' and a Z'. Both new bosons are left-handed and generally taken to be approximately equal in mass. There are two parameters: a mixing angle ϕ, which represents a mixing between the charged bosons of the two $SU(2)$ symmetries, and $x = (u/v)^2$, where u and v are the VEV's of the two scalar multiplets of the model. The relation $M_{Z'} \simeq M_{W'}$ follows in the limit $x/\sin^2\phi \gg 1$ and the parameter x may be replaced by $M_{W'}$, so that only ϕ enters as a parameter

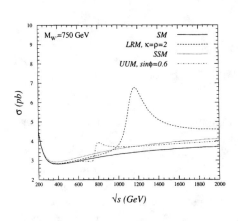

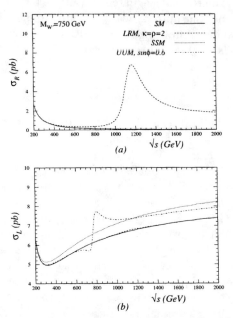

FIGURE 2. Unpalarized total cross section versus center-of-mass energy in the SM, LRM, SSM and UUM.

FIGURE 3. As Fig. 2, except for (a) right-handed; (b) left-handed e^- beam.

for determining mass discovery limits. The leptonic couplings may be inferred from the Lagrangian

$$\mathcal{L}_{\text{lept}} = -\frac{g \sin\phi}{2\cos\phi}[\sqrt{2}\,\overline{\psi}_{\nu_l}\gamma_\mu\psi_{l,L}W_2^{+,\mu} + (\overline{\psi}_{\nu_l}\gamma_\mu\psi_{\nu_l,L} - \overline{\psi}_l\gamma_\mu\psi_{l,L})Z_2^\mu], \tag{5}$$

where $\psi_L = \frac{1}{2}(1-\gamma_5)\psi$. The existing constraint on ϕ is [1]

$$0.24 \lesssim \sin\phi \lesssim 0.99 \tag{6}$$

CROSS SECTIONS

As inputs, we take $M_W = 80.33$ GeV, $M_Z = 91.187$ GeV, $\sin^2\theta_W = 0.23124$, $\alpha = 1/128$, $\Gamma_Z = 2.49$ GeV. Let E_γ, θ_γ denote the photon's energy and angle in the e^+e^- center-of-mass, respectively. No binning or transverse momentum cuts have been explicitly introduced at this point. However, we have restricted the range of E_γ, θ_γ as follows:

$$E_\gamma > 10 \text{ GeV}, \quad 10^\circ < \theta_\gamma < 170^\circ, \tag{7}$$

so that the photon may be detected cleanly. As well, the angular cut eliminates the collinear singularity arising when the photon is emitted parallel to the beam.

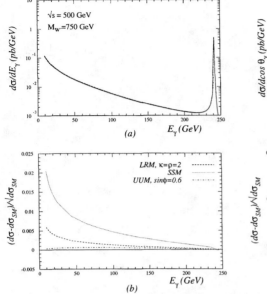

FIGURE 4. (a) Cross section and (b) relative statistical significance of the deviation from the SM versus photon energy.

FIGURE 5. (a) Cross section and (b) relative statistical significance of the deviation from the SM versus $\cos\theta_\gamma$.

Figure 2 shows the total unpolarized cross section versus center-of-mass energy for the SM, LRM (taking $\kappa = \rho = 2$), SSM and UUM (taking a representative value of $\sin\phi = 0.6$). Throughout, we use $M_{W'} = 750$ GeV. Figure 3 shows the same, except for pure right- and left-handed e^- beams. The peaks are due to the Z's in the LRM and UUM. As expected, the LRM gives a large effect when the e^- is right-handed while the SSM and UUM give a larger effect for a left-handed e^-. The fact that the SM right-handed cross section goes to zero for large $\sqrt{s}$ indicates W (t-channel) dominance well above the Z pole.

In Figure 4(a), we plot the differential cross section with respect to E_γ for $\sqrt{s}$ of 500 GeV. The peak in the photon distributions is due to the radiative return to the Z resonance. At higher energies, there are additional peaks in the E_γ spectrum due to Z's. We see that most of the contribution comes from the lower E_γ range. This must be weighted with the deviation from the SM in order to gauge the relative statistical significance of the various energy regions. This is done in Figure 4(b) where $(d\sigma/dE_\gamma - d\sigma_{SM}/dE_\gamma)/\sqrt{d\sigma_{SM}/dE_\gamma}$ is plotted as a function of E_γ. Indeed, we see that one benefits little from the region E_γ above ~ 200 GeV in all models.

In Figure 5 we plot the the differential cross section with respect to $\cos\theta_\gamma$ and the corresponding relative statistical significance. We see that both the cross section and relative statistical significance are peaked in the forward/backward direction and the distributions are very nearly symmetric in $\cos\theta_\gamma$. In Figures 4(b) and 5(b),

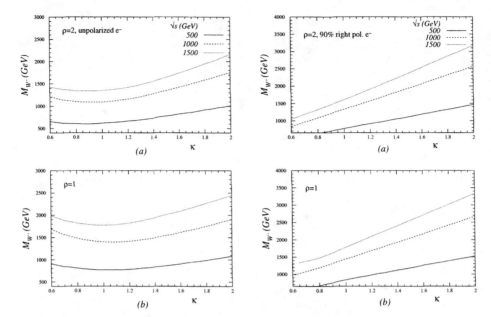

FIGURE 6. W' 95% C.L. discovery limits versus κ in the LRM obtained with unpolarized beam for (a) $\rho = 2$; (b) $\rho = 1$.

FIGURE 7. As Fig. 6, except with 90 percent right-polarized electron beam.

the overall normalization is unimportant. Experimentally, some binning scheme will be adopted and each bin will carry a weight proportional to the beam luminosity.

NLC W' MASS DISCOVERY LIMITS

For $\sqrt{s}$ of 500 GeV, we take an integrated luminosity of $50 fb^{-1}$ and for $\sqrt{s}$ of 1 TeV and 1.5 TeV we take $200 fb^{-1}$. For the polarized limits we take half the above luminosities assuming equal running in both polarization states (of the e^- beam). We assume 90% e^- polarization unless otherwise stated.

In obtaining limits, we have imposed the additional cut:

$$E_\gamma < E_{\gamma,\text{max}} , \qquad (8)$$

in order to cut out the high energy events, especially those near the Z pole, which are insensitive to W's and Z's. It was found that $E_{\gamma,\text{max}}$ of 200, 350 and 500 GeV for $\sqrt{s}$ of 500 GeV, 1 TeV and 1.5 TeV, respectively, lead to the best limits in general, although the limits were not very sensitive to moderate variations in $E_{\gamma,\text{max}}$. The limits are given at 95% confidence level and are calculated for all three energies.

Figure 6 presents the W' mass discovery limits obtainable with an unpolarized beam for the LRM, plotted versus κ for $\rho = 1, 2$. Depending on $\sqrt{s}$, ρ and κ, they range from 600 GeV to 2.5 TeV. The predicted dependence on κ and ρ is

TABLE 1. W' 95% C.L. discovery limits obtained in the SSM, LRM ($\kappa = \rho = 2$) and the UUM ($\sin\phi = 0.6$), assuming 90 percent e^- polarization (unless otherwise stated) and using 1/2 the unpolarized luminosity for the left and right cases.

$\sqrt{s}$ (GeV)	Model	Unpolarized e^- Limit (TeV)	Left Pol. e^- Limit (TeV)	Right Pol. e^- Limit (TeV)	100% L/R Pol. Limit (TeV)
500	SSM	2.45	2.45	1.15	2.45
(50 fb^{-1})	LRM	1.0	<0.75	1.45	2.05
	UUM	0.65	0.65	0.55	0.65
1000	SSM	4.55	4.5	2.15	4.55
(200 fb^{-1})	LRM	1.75	<1	2.55	4.5
	UUM	1.3	1.3	1.15	1.3
1500	SSM	5.2	5.15	2.45	5.2
(200 fb^{-1})	LRM	2.15	<1.25	3.2	6.2
	UUM	1.85	1.85	1.65	1.85

generally observed, except at low κ where we notice a moderate *increase* in the limits, even though the W' couplings have weakened. We attribute this effect to the Z', whose couplings are enhanced (but its mass increased) in the low κ region. This is evidenced by the appreciable improvement in the bounds for low κ and $\rho = 1$. Figure 7 demonstrates the improvement in bounds in the moderate to large κ region obtained when a (90%) polarized right-handed e^- beam is used. The beam polarization picks out the LRM W' and suppresses the SM W. Increase of the polarization can lead to even higher limits as shown in the rightmost column of Table 1, where limits are tabulated for all three models.

The highest limits are obtained for the SSM in most scenarios, except when $\sin\phi$ is large, as indicated in Figure 8 which shows the limits in the UUM versus $\sin\phi$. We observe a turn-on in sensitivity for $\sin\phi \gtrsim 0.62$ at $\sqrt{s} = 500$ GeV and 1 TeV, while for $\sqrt{s} = 1.5$ TeV this occurs for $\sin\phi \gtrsim 0.73$. The interference term may play a role in this behaviour. From another perspective, for fixed $\sin\phi$, one may observe sudden changes in sensitivity as $\sqrt{s}$ is varied as can be seen from the changing of the sign of the effect on the cross section in Figs 2,3. The result is that for $0.62 \lesssim \sin\phi \lesssim 0.72$, we obtain better limits at $\sqrt{s} = 1$ TeV than we do at $\sqrt{s} = 1.5$ TeV.

For both the UUM and the SSM, where the W's are left-handed, there is little benefit from polarization. The reason is that all the effect comes from the left-handed e^- initial state, which also dominates the unpolarized cross section. After folding in the luminosity decrease associated with running in a particular polarization state, all benefits are lost.

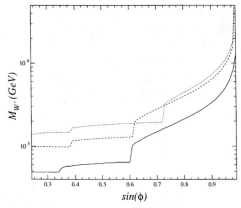

FIGURE 8. 95% C.L. discovery limits versus $\sin(\phi)$ in the UUM; lines as in Fig. 6.

SUMMARY AND OUTLOOK

The usefulness of the process $e^+e^- \to \nu\bar{\nu} + \gamma$ in searching for W''s has been demonstrated and should be complimentary to direct searches at the LHC. The results of our study will be extended to include cuts on the transverse momentum of the photon to reduce backgrounds (primarily from radiative Bhabha scattering with undetected e^+e^-) and to examine the effect of binning. All remaining backgrounds will have to be included in the final analysis of the data and are currently under investigation, but are not expected to significantly affect our limits. Other models are also under consideration.

REFERENCES

1. Barger, V., and Rizzo, T., *Phys. Rev.* **D41**, 946 (1990).
2. Hewett, J., SLAC-PUB-7441, June 1996, hep-ph/9704292.
3. Particle Data Group (Caso, C., et al.), *Eur. Phys. J.* **C3**, 1 (1998).
4. Barenboim, G., Bernabéu, J., Prades, J., and Raidal, M., *Phys. Rev.* **D55**, 4213 (1997).
5. For a review, see: Cvetič, M., and Godfrey, S., "Discovery and Identification of Extra Gauge Bosons," in *Electro-weak Symmetry Breaking and Beyond the Standard Model*, eds. Barklow, T., et al., World Scientific, 1995, hep-ph/9504216.
6. Mohapatra, R.N., *Unification and Supersymmetry*, New York: Springer, 1986, and original references therein.
7. Parida, M., and Raychaudhuri, A., *Phys. Rev.* **D26**, 2364 (1982); Chang, D., Mohapatra, R., and Parida, M., *Phys. Rev.* **D30**, 1052 (1984).
8. Georgi, H., Jenkins, E.E., and Simmons, E.H., *Phys. Rev. Lett.* **62**, 2789 (1989); **63**, 1540(E) (1989).

Closing the Low-mass Axigluon Window

Michael A. Doncheski

Department of Physics
Penn State University
Mont Alto, PA 17237

Abstract. In this report, I will present the current status of the low-mass axigluon. The axigluon is a massive, color octet, axial vector boson, predicted in, *e.g.*, chiral color models and some technicolor models, with a mass of order the electroweak scale. Axigluons with a mass larger than about 125 GeV to nearly 1 TeV can be eliminated by di-jet production at hadron colliders like the TEVATRON, but a low-mass window exists that the di-jet search can not probe. Υ decays can rule out axigluons with a mass up to 25 GeV, and low energy e^+e^- (PEP and PETRA) can rule out axigluons with a mass up to 50 GeV using a measurement of R. Top production at the TEVATRON disfavors a light axigluon. A measurement of R at LEP **strongly** disfavors a light axigluon, and rules out an axigluon with mass $<$ 365 GeV.

MOTIVATION

The possible existence of an axigluon was first realized in chiral color models [1], where the gauge group of the strong interaction is extended from $SU(3)_C$ to $SU(3)_L \times SU(3)_R$. At low energy, this larger color gauge group breaks to the usual $SU(3)_C$ with its octet of massless vector gluons, but it leaves a residual $SU(3)$ with an octet of massive, axial vector bosons called *axigluons*. In these chiral color models, the axigluon is expected to have a mass of order the electroweak scale. Similar states are predicted in technicolor models [2].

In order to search for these states, Bagger, Schmidt and King [3] noted that the di-jet cross section at hadron colliders would be modified by the addition of s-channel axigluon exchange. Searches were performed by the UA1 and CDF collaborations, with limits of 150 $GeV < M_A <$ 310 GeV by UA1 [4] and 120 $GeV < M_A <$ 980 GeV by CDF [5]. Given additional center of mass energy and/or luminosity, these di-jet searched at hadron colliders will easily raise the upper exclusion limit, but it will be difficult to decrease the lower exclusion limit.

Several additional search strategies were suggested involving the Z^0 and the large amounts of data taken by the LEP experiments. Rizzo [6] suggested $Z^0 \to q\bar{q}A$ and Carlson, *et al.*, [7] suggested $Z^0 \to gA$ going through a quark loop. These suggestions involve low rates, and the former requires precision multi-jet reconstruction.

In the remainder of this report, I will address some additional search strategies for the axigluon, and report on the current status of the search.

Υ DECAYS TO REAL AXIGLUONS

The decay of the Υ family is an ideal area to search for low mass, strongly interacting particles. In the Standard Model, the dominant hadronic decay mode of any heavy vector quarkonium state ($J^{PC} = 1^{--}$) is the 3 gluon mode, $V_Q \to ggg$, where Q refers to the specific flavor of heavy quark. The decay to a single gluon is forbidden by color, while the decay to 2 gluons is forbidden by both the Landau-Pomeranchuk-Yang theorem [8] (which forbids the decay of a $J = 1$ state to 2 massless spin 1 states) and quantum numbers ($C = -1$ for the gluon, so an odd number of gluons are needed for this particular decay). The leading order decay rate of a heavy quarkonium state to 3 gluons is well known:

$$\Gamma(V_Q \to ggg) = \frac{40(\pi^2 - 9)\alpha_s^3}{81\pi M_V^2}|R(0)|^2 \tag{1}$$

where M_V is the quarkonium state's mass and R(0) is the non-relativistic, radial wavefunction evaluated at the origin.

A heavy, vector quarkonium state **may** decay into a gluon plus an axigluon. As the axigluon is massive, the Landau-Pomeranchuk-Yang theorem is avoided, and the axigluon has $C = +1$. The decay rate for $V_Q \to Ag$ is given by [9]:

$$\Gamma(V_Q \to Ag) = \frac{16\alpha_s^2}{9M_V^2}|R(0)|^2(1-x)(1+\frac{1}{x}) \tag{2}$$

where $x = \left(\frac{M_A}{M_V}\right)^2$. Both this decay rate and the leading order Standard Model rate depend on the non-relativistic radial wavefunction; a ratio of these two decay rates does not depend on the wavefunction, and, as such, had much less uncertainty. The ratio is given by:

$$\frac{\Gamma(V_Q \to Ag)}{\Gamma(V_Q \to ggg)} = \frac{18\pi}{5\alpha_s(\pi^2 - 9)}(1-x)(1+\frac{1}{x}) \tag{3}$$

Notice that, since the gluon plus axigluon mode has one fewer power of α_s, the ratio is large (the numerical factor in front of the kinematical structure is approximately 100).

This ratio, as a function of x is shown in Figure 1. The addition of this new hadronic decay mode will at least double the hadronic width of a vector quarkonium state, even for an axigluon mass nearly equal to the quarkonium state mass. Using this process and the Υ system, we can exclude an axigluon with mass below about 10 GeV. A analysis by Cuypers and Frampton [10] yielded quantitatively similar conclusions.

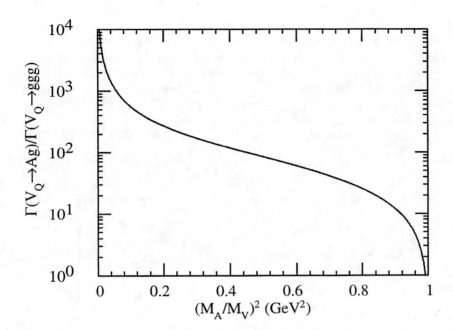

FIGURE 1. Ratio $\dfrac{\Gamma(V_Q \to Ag)}{\Gamma(V_Q \to ggg)}$ for the decay of a heavy quarkonium state to a real axigluon.

Υ DECAYS TO VIRTUAL AXIGLUONS

In addition to Υ decays to real axigluons, it is possible to study Υ decays to virtual axigluons, $\Upsilon \to gA^*(\to q\bar{q})$. The decay rate is given by [11]

$$\Gamma(V_Q \to q\bar{q}g) = \frac{2^8 n \alpha_s^3}{3^5 \pi} \frac{M_V^2}{M_A^4} F(x) |R(0)|^2 \quad (4)$$

where n is the number of active quark flavors (in this case 4) and

$$F(x) = \frac{3}{2} x^2 \left(2x \ln\left(\frac{x}{x-1}\right) - 2 - \frac{1}{x} \right). \quad (5)$$

As before, we can look at the ratio of this hadronic width to the dominant Standard Model width:

$$\frac{\Gamma(V_Q \to q\bar{q}g)}{\Gamma(V_Q \to ggg)} = \frac{128 F(x)}{15(\pi^2 - 9)x^2}. \quad (6)$$

This time, there is no large numerical factor. This ratio is shown in Figure 2. The dashed lines in the figure indicate 2 possible exclusion limits that can be made using data. The more conservative estimate is to argue that our knowledge of the Υ width is such that a correction to the standard width larger than 50% is unacceptable; thus, this ratio is smaller than 0.5, which gives an upper exclusion limit of $M_A < 21\ GeV$. A less conservative estimate is to compare this correction to the expected rate to QCD radiative corrections to the Standard Model rate and other possible contributions to the hadronic width (e.g., $\Upsilon \to \gamma^* \to q\bar{q}$), and argue that another correction larger than these is unacceptable. In this case, the ratio must be less than 0.25, excluding axigluons with mass smaller than $25\ GeV$.

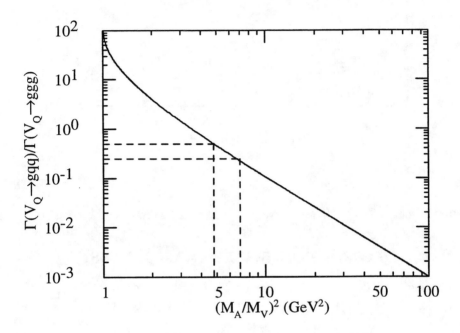

FIGURE 2. Ratio $\frac{\Gamma(V_Q \to q\bar{q}A)}{\Gamma(V_Q \to ggg)}$ for the decay of a heavy quarkonium state to a virtual axigluon. The dashed lines indicate 2 possible exclusion limits.

Not long after our work on the Υ, Cuypers and Frampton and Cuypers, Falk and Frampton [12] published papers on the R value in e^+e^- collisions at low energy. They included the full set of QCD radiative corrections, including axigluon radiative corrections, to the tree level process. They exclude an axigluon with $M_A < 50\ GeV$ using PEP and PETRA data.

TOP PRODUCTION

The top is too short lived to allow for a toponium state; if it did, the same techniques that worked in the Υ system would work for toponium as well. On the other hand, because of the large mass of the top, top production is inherently perturbative, $q\bar{q} \to t\bar{t}$ is well understood, and it can be used to search for a light axigluon. The parton level cross section for $q\bar{q} \to t\bar{t}$, due to an s-channel gluon, is well known:

$$\left(\frac{d\sigma}{d\hat{t}}\right)_0 = \frac{1}{16\pi\hat{s}^2} \frac{64\pi^2}{9} \alpha_s^2 \left[\frac{(m^2 - \hat{t})^2 + (m^2 - \hat{u})^2 + 2m^2\hat{s}}{\hat{s}^2}\right] \quad (7)$$

and the cross section with the addition of an s-channel axigluon is [13]

$$\left(\frac{d\sigma}{d\hat{t}}\right)_{q\bar{q}} = \left(\frac{d\sigma}{d\hat{t}}\right)_0 \left[1 + |r(\hat{s})|^2 + 4\Re(r(\hat{s}))\frac{(\hat{t} - \hat{u})\hat{s}\beta}{(\hat{t} - \hat{u})^2 + \hat{s}^2\beta^2}\right] \quad (8)$$

where $r(\hat{s}) = \frac{\hat{s}}{\hat{s} - M_A^2 + iM_A\Gamma_A}$ and β is the top quark velocity parameter, $\beta = \sqrt{1 - \frac{4m^2}{\hat{s}}}$. The addition of an s-channel axigluon affects both the total cross section and the forward-backward asymmetry (only the interference term affects the forward-backward asymmetry).

The results on total cross section are shown in Figure 3. From the relatively good agreement between experimental values of the top cross section [14,15] and theoretical calculations [16,17], we can say that an axigluon is disfavored by top production cross section, but nothing conclusive can be said.

Shown in Figure 4 is the forward-backward asymmetry in top production as a function of axigluon mass. Without an axigluon, the asymmetry is identically zero.

MISCELLANEOUS

Unitarity is violated, in that $Q\bar{Q} \to Q\bar{Q}$ will be non-perturbative unless

$$M_A > \sqrt{\frac{5\alpha_s}{3}} M_Q \quad (9)$$

as pointed out by Robinett [18]. Using the top quark as Q, this leads to a lower limit on the axigluon mass of $M_A > 73\ GeV$ [19].

Higgs searches, e.g., by CDF, can make use of the process $p\bar{p} \to W + X^0$, where X^0 is the neutral Higgs boson, and it is assumed to decay to $b\bar{b}$ [20]. The limit on Higgs boson mass is such that $\sigma \cdot BR > 15 - 20\ pb$ are not allowed. The same final state is possible with an axigluon in place of the Higgs boson; we find the part level cross section for $q\bar{q}' \to WA$ to be [19]:

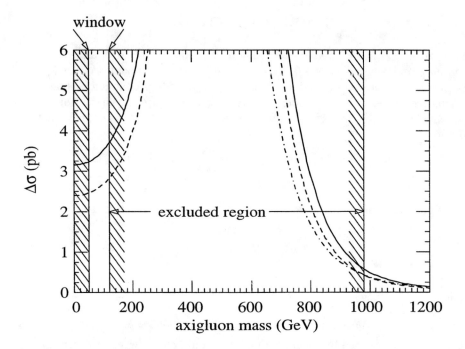

FIGURE 3. Difference in top production cross section, based on the presence of an axigluon. The different curves are for different sets of leading order parton distribution functions, and different choices for axigluon width. The solid (dotdashed) line is for the "new" Duke and Owens pdf's with $\Gamma_A = 0.1 M_A$ ($0.2 M_A$); the dashed line is for CTEQ4L with $\Gamma_A = 0.1 M_A$.

$$\frac{d\hat{\sigma}}{d\hat{t}} = \frac{4\alpha_s}{9} \left[\frac{G_F M_W^2}{\sqrt{2}} \right] \frac{|V_{qq'}|^2}{\hat{u}\hat{t}\hat{s}^2} \left[\hat{u}^2 + \hat{t}^2 + 2\hat{s}(M_W^2 + M_A^2) - \frac{M_A^2 M_W^2 (\hat{u}^2 + \hat{t}^2)}{\hat{u}\hat{t}} \right]. \quad (10)$$

Assuming $BR(A \to b\bar{b}) = \frac{1}{5}$, and calculating the cross section for the associated production of $W + A$, a conservative lower limit of $M_A > 70\ GeV$ is possible, using the same analysis at the Higgs search.

Finally, we can examine the value of α_s, extracted from low energy data but run up to M_Z to the value of α_s extracted from the hadronic width of the Z^0 at the pole. Since the axigluon mass is expected to be at least $70\ GeV$, the running of α_s should not be affected much by the axigluon. Then, the R value at low energy, or the hadronic width at the Z^0 pole, is subject to a correction from real and virtual axigluons [12], of the form:

$$\left[1 + \frac{\alpha_s(\sqrt{s})}{\pi} f\left(\frac{\sqrt{s}}{M_A}\right) + \mathcal{O}(\alpha_s^2) \right] \quad (11)$$

where the function $F(\sqrt{s}/M_A)$ is calculated in Ref. [12]. The Particle Data

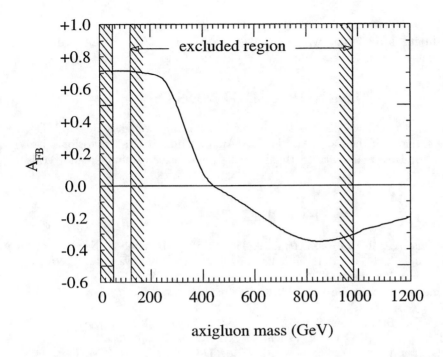

FIGURE 4. Top production forward-backward asymmetry.

Group [21] quotes a value of α_s from various low energy data run up to M_Z as $\alpha_s^{(LE)} = 0.118 \pm 0.004$, while the extraction from the hadronic wdt of the Z^0 at M_Z as $\alpha_s^{(HE)} = 0.123 \pm 0.004 \pm 0.002$. Attributing the difference in the extracted values of α_s to the axigluon gives abound on the $f(\sqrt{s}/M_A)$ term, such that $f(M_Z/M_A) \leq 0.042 \pm 0.050$. This implies that $f(M_Z/M_A) < 0.092(0.142)$ at the 65% (95%) level, and that $M_Z > 570\ GeV(365\ GeV)$ at the same confidence levels. Should the agreement between the low energy and high energy extractions of α_s increase, the corresponding lower limit on the axigluon mass would also increase.

CONCLUSIONS

The existence of an axigluon is predicted in chiral color models. A low-mass axigluon is difficult to exclude in typical collider experiments (*e.g.*, using di-jet data). Other approaches must be used to rule out axigluons with masses below 125 GeV. Υ decay, top production, unitarity bounds and associated production of a W boson with an axigluon can exclude axigluons with mass below about 70 GeV. A comparison of α_s as extracted in low energy experiments and high energy experiments can rule out an axigluon with a mass lower than 365 GeV.

This completely closes the low-mass axigluon window, and when combined with the CDF limits, an axigluon with mass below about 1 TeV is not allowed. An axigluon, if it exists, is in the realm of TeV physics.

ACKNOWLEDGEMENTS

I would like to acknowledge the support of Penn State through a Research Development Grant and the Mont Alto Faculty Affairs Committee's Professional Development Fund. I would like to thank Rick Robinett for carefully reading this manuscript.

REFERENCES

1. J. Pati and A. Salam, Phys. Lett. **B58**, 333 (1975); L. Hall and A. Nelson, Phys. Lett. **B153**, 430 (1985); P. H. Frampton and S. L. Glashow, Phys. Lett. **B190**, 157 (1987); Phys. Rev. Lett. **58**, 2168 (1987).
2. S. Dimopoulos, Nucl. Phys. **B168**, 69 (1980); J. Preskill, Nucl. Phys. **B177**, 21 (1981).
3. J. Bagger, C. Schmidt and S. King, Phys. Rev. **D37**, 1188 (1988).
4. C. Albajar, et al. (UA1 Collaboration), Phys. Lett. **B209**, 127 (1988).
5. F. Abe, et al. (CDF Collaboration), Phys. Rev. **D41**, 1722 (1990); Phys. Rev. Lett. **74**, 3538 (1995); Phys. Rev. **D55**, R5263 (1997).
6. T. G. Rizzo, Phys. Lett. **B197**, 273 (1987).
7. E. D. Carlson, S. L. Glashow and E. Jenkins, Phys. Lett. **B202**, 281 (1988).
8. L. D. Landau, Dokl. Akad. Nauk SSSR **60**, 207 (1948); I. Ya. Pomeranchuk, Dokl. Akad. Nauk SSSR **60**, 263 (1948); C. N. Yang, Phys. Rev. **77**, 55 (1950).
9. M. A. Doncheski, H. Grotch and R. Robinett, Phys. Lett. **B206**, 137 (1988).
10. F. Cuypers and P. H. Frampton, Phys. Rev. Lett. **60**, 1237 (1988).
11. M. A. Doncheski, H. Grotch and R. W. Robinett, Phys. Rev. **D38**, R412 (1988).
12. F. Cuypers and P. H. Frampton, Phys. Rev. Lett. **63**, 125 (1989); A. Falk, Phys. Lett. **B230**, 119 (1989); F. Cuypers, A. F. Falk and P. H. Frampton, Phys. Lett. **B259**, 173 (1991).
13. M. A. Doncheski and R. W. Robinett, Phys. Lett. **B412**, 91 (1997).
14. B. Abbott, et al. (D0 Collaboration), FERMILAB-PUB-99-008-E (hep-ex/9901023).
15. G. V. Velev (for the CDF Collaboration), FERMILAB-CONF-98-192-E.
16. E. L. Berger and H. Contopanagos, Phys. Rev. **D54**, 3085 (1996).
17. S. Catani, et al., Phys. Lett. **B378**, 329 (1996).
18. R. W. Robinett, Phys. Rev. **D39**, 834 (1989).
19. M. A. Doncheski and R. W. Robinett, Phys. Rev. **D58**, 097702-1.
20. P. Bhat, invited talk at PHENO-CTEQ-98: Frontiers of Phenomenology from Non-Perturbative QCD to New Physics, Madison, 1998.
21. R. M. Barnett, et al. (Particle Data Group), Phys. Rev. **D54**, 1 (1996).

Lowest-Lying Scalar Mesons and a Possible Probe of Their Quark Substructure

Amir H. Fariborz

Department of Physics, Syracuse University, Syracuse, New York 13244-1130, USA.

Abstract. In this talk, an overview of the status of the light scalar mesons in the context of the non linear chiral Lagrangian of references [1–3] is presented. The evidence for the existence of a scalar nonet below 1 GeV is reviewed, and it is shown that by introducing a scalar nonet an indirect way of probing the quark substructure of these scalars through the scalar mixing angle can be obtained. It is then reviewed that consistency of this non-linear chiral Lagrangian framework with the experimental data on $\pi\pi$ and πK scattering, as well as the decay $\eta' \to \eta\pi\pi$, results in a range for the mixing angle which indicates that the quark substructure of these light scalars are closer to a four quark picture.

I INTRODUCTION

Lowest-lying scalar mesons (scalar mesons below 1 GeV) are of fundamental importance in understanding the theory and phenomenology of low energy QCD. However, the properties of these scalars, in particular their quark substructure, are not quite understood. As a result they are at the focus of many theoretical and experimental investigations.

From the experimental point of view, there are at least four well established light scalars – the isosinglet $f_0(980)$ and the isotriplet $a_0(980)$. There are also five other candidates – the isosinglet $\sigma(560)$, and two isodoublets $\kappa(900)$ and $\bar{\kappa}(900)$, which are not quite established experimentally. In 1998 edition of Particle Date Group [4], the $\sigma(560)$ is listed as $f_0(400-1200)$ with a very uncertain properties; a mass between 400 to 1200 MeV and a very broad decay width between 600 to 1000 MeV. In a recent experimental study of τ lepton decay by CLEO collaboration [5], a significant contribution due to the σ is pointed out and it is reported that inclusion of a σ with $m_\sigma = 555$ MeV and $\Gamma_\sigma = 540$ MeV significantly improves the fits. The situation of $\kappa(900)$ is not clear experimentally. Therefore, altogether there are 9 possible candidates for lowest lying scalar mesons.

The situation of these scalars is also not clear from the theoretical point of view; there are model dependent calculations, and as a result, different conclusions.

However, within any theoretical framework two basic questions should be addressed:
1. Is there a clear evidence for the existence of a $\sigma(560)$ and a $\kappa(900)$?
2. Can we describe the properties of these mesons like their masses, decay widths, interactions, and in particular their quark substructure?

In this talk we first review the general nonlinear chiral Lagrangian framework of refs. [1–3], and discuss how within this framework a $\sigma(560)$ was observed in [6,7], and a $\kappa(900)$ was observed in [1]. We then review how this nonlinear chiral Lagrangian can be rewritten in terms of a scalar nonet by introducing a few new free parameters. We show how these parameters as well as the acceptable range of the scalar mixing angle can be fixed by considering the $\pi\pi$ and πK scattering, and the $\eta' \to \eta\pi\pi$ decay. Based on the acceptable range of the scalar mixing angle we discuss that the quark substructure of the light scalar mesons is closer to a four quark picture. We conclude by summarizing the results.

II OUR THEORETICAL FRAMEWORK AND EVIDENCE FOR THE $\sigma(560)$ AND $\kappa(900)$

We work within the effective non-linear chiral Lagrangian framework. The pseudoscalar part of the Lagrangian is [1]

$$\mathcal{L}_\phi = -\frac{F_\pi^2}{8}\mathrm{Tr}\left(\partial_\mu U \partial_\mu U^\dagger\right) + \mathrm{Tr}\left[\mathcal{B}\left(U + U^\dagger\right)\right] \quad (1)$$

with $F_\pi = 131$ MeV, $\mathcal{B} = m_\pi^2 F_\pi^2/8\,\mathrm{diag}(1,1,2m_K^2/m_\pi^2-1)$ is the dominant symmetry breaking term, and $U = e^{2i\frac{\phi}{F_\pi}} = \xi^2$ where ϕ is the pseudoscalar nonet

$$\phi_a^b = \begin{bmatrix} \frac{\eta_{NS}+\pi_0^0}{\sqrt{2}} & \pi^+ & K^+ \\ \pi^- & \frac{\eta_{NS}-\pi_0^0}{\sqrt{2}} & K^0 \\ K^- & \bar{K}^0 & \eta_S \end{bmatrix}. \quad (2)$$

U transforms linearly under chiral transformation [$U \to U_L U U_R^\dagger$ with $U_{L,R} \in U(3)_{L,R}$], whereas ξ transforms nonlinearly [$\xi \to U_L \xi K^\dagger(\phi,U_L,U_R) = K(\phi,U_L,U_R)\xi U_R^\dagger$].

The vectors can be introduced in this framework in terms of the vector nonet ρ with a Lagrangian that has the same form as that of usual gauge fields

$$\mathcal{L}_\rho = -\frac{1}{2}m_v^2 \mathrm{Tr}\left[\left(\rho_\mu - \frac{v_\mu}{\tilde{g}}\right)^2\right] - \frac{1}{4}\mathrm{Tr}\left[F_{\mu\nu}(\rho)F_{\mu\nu}(\rho)\right] \quad (3)$$

with $F_{\mu\nu} = \partial_\mu \rho_\nu - \partial_\nu \rho_\mu - i\tilde{g}[\rho_\mu,\rho_\nu]$. In (3), $p_\mu = \frac{i}{2}\left(\xi\partial_\mu\xi^\dagger - \xi^\dagger\partial_\mu\xi\right)$ and $v_\mu = \frac{i}{2}\left(\xi\partial_\mu\xi^\dagger + \xi^\dagger\partial_\mu\xi\right)$, and have simple transformation properties under chiral transformation.

We introduce scalars into this picture in two stages. First in order to see whether there is an indication of $\sigma(560)$ and $\kappa(900)$ within our framework, we introduce scalars in a phenomenological way. We consider a general isospin invariant form [1]

$$\mathcal{L}_s = -\frac{\gamma_{\sigma\pi\pi}}{\sqrt{2}}\sigma\partial_\mu\pi\cdot\partial_\mu\pi - \frac{\gamma_{\sigma K\bar{K}}}{\sqrt{2}}\sigma\left(\partial_\mu K^+\partial_\mu K^- + \cdots\right) - \frac{\gamma_{f_0\pi\pi}}{\sqrt{2}}f_0\partial_\mu\pi\cdot\partial_\mu\pi$$
$$-\frac{\gamma_{f_0 K\bar{K}}}{\sqrt{2}}f_0\left(\partial_\mu K^+\partial_\mu K^- + \cdots\right) - \gamma_{\kappa K\pi}\left(\kappa^0\partial_\mu K^-\partial_\mu\pi^+ + \cdots\right) \quad (4)$$

and take the coupling constants as independent parameters – we either take the couplings as fitting parameters or input them from experimental measurements.

Now in order to see whether this framework sees a σ and/or a κ, we can consider computing processes to which these mesons could significantly contribute, and then compare our prediction to the experimental data and search for signs of these scalars. In principle $\sigma(560)$ and $\kappa(900)$ can be probed in $\pi\pi$ and πK scattering, respectively. In fact, their contributions could be substantial as they appear as poles in the scattering amplitudes. To see the effect of κ, using Lagrangian (4), appropriate πK scattering amplitudes were computed in [1] and the result were matched to the experimental data. The scattering amplitudes were computed by only taking tree level Feynman diagrams into account – this is motivated by $1/N_c$ expansion. It was shown in [1] that a κ with a mass around 900 MeV and a decay width around 320 MeV is needed in order to describe the experimental data on the πK scattering. This technique was first developed in [6,7], in which it was shown that in order to agree with the $\pi\pi$ experimental data there is a need for a σ meson with a mass around 550 MeV and a decay width around 370 MeV.

Therefore within our theoretical framework there is a clear signal for both the $\sigma(560)$ and the $\kappa(900)$. This motivates us, in our second stage of investigation, to combine these scalars together with the $f_0(980)$ and the $a_0(980)$ into a light scalar nonet, in terms of which we rewrite our Lagrangian in the next section.

III A POSSIBLE SCALAR NONET BELOW 1 GEV

We now construct a light scalar nonet out of $\sigma(560)$, $\kappa(900)$, $f_0(980)$ and $a_0(980)$ in the form

$$N = \begin{bmatrix} N_1^1 & a_0^+ & \kappa^+ \\ a_0^- & N_2^2 & \kappa^0 \\ \kappa^- & \bar{\kappa}^0 & N_3^3 \end{bmatrix}. \quad (5)$$

In general, $\sigma(560)$ and $f_0(980)$ are a mixture of $(N_1^1+N_2^2)/\sqrt{2}$ and N_3^3. We represent this mixing in terms of a scalar mixing angle θ_s as

$$\begin{pmatrix} \sigma \\ f_0 \end{pmatrix} = \begin{pmatrix} \cos\theta_s & -\sin\theta_s \\ \sin\theta_s & \cos\theta_s \end{pmatrix} \begin{pmatrix} N_3^3 \\ \frac{N_1^1+N_2^2}{\sqrt{2}} \end{pmatrix}. \quad (6)$$

In this parametrization, $\theta_s = \pm\pi/2$ corresponds to the conventional ideal mixing when a pure $q\bar{q}$ assignment is used for N. Another interesting limit is $\theta_s = 0$ which corresponds to the dual ideal mixing when a pure four-quark assignment is used to describe N [8].

We can now rewrite the scalar sector of our Lagrangian in terms of the nonet (5)

$$\mathcal{L}_{mass} = -a\text{Tr}(NN) - b\text{Tr}(NN\mathcal{M}) - c\text{Tr}(N)\text{Tr}(N) - d\text{Tr}(N)\text{Tr}(N\mathcal{M})$$
$$\mathcal{L}_{N\phi\phi} = A\epsilon^{abc}\epsilon_{def}N_a^d\partial_\mu\phi_b^e\partial_\mu\phi_c^f + B\text{Tr}(N)\text{Tr}(\partial_\mu\phi\partial_\mu\phi)$$
$$+C\text{Tr}(N\partial_\mu\phi)\text{Tr}(\partial_\mu\phi) + D\text{Tr}(N)\text{Tr}(\partial_\mu\phi)\text{Tr}(\partial_\mu\phi) \qquad (7)$$

with $\mathcal{M} = \text{diag}(1,1,x)$ is the spurion matrix with x the ratio of strange to non-strange quark masses. The mass part of the Lagrangian is given in terms of new free parameters a, b, c, d, and θ_s which can be determined by inputting the scalar masses $m_\sigma, m_{f0}, m_\kappa$ and m_{a0}. We find that our model restricts m_κ in the range 685 to 980 MeV, and also for any input of scalar masses there are two solutions for θ_s. Thus, in our framework there are two acceptable ranges for the scalar mixing angle which are shown in Fig. 1. These are the *large angle solution*: $36° \leq \theta_s \leq 90°$ and $-90° \leq \theta_s \leq -71°$, and the *small angle solution*: $-71° \leq \theta_s \leq 36°$. As m_κ varies from its minimum value to its maximum, these regions are entirely swept through. There are also new free parameters A, B, C, and D in the scalar-pseudoscalar-pseudoscalar interaction part of Lagrangian which can be determined

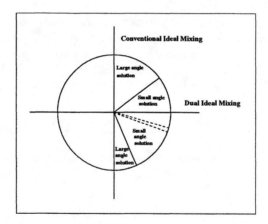

FIGURE 1. Two regions for the scalar mixing angle for the acceptable range of 685 MeV$\leq m_\kappa \leq$ 980 MeV. As m_κ varies from its minimum value to its maximum value, θ_s in the small angle region varies from 36° to -71°, and in the large angle region θ_s varies from 36° to 90°, then to -90° and finally to -71°. The region bounded between dashed lines ($-20° \leq m_\kappa \leq -15°$) corresponds to 875 MeV $\leq m_\kappa \leq$ 897 MeV, and is consistent with experimental data on $\pi\pi$ and πK scattering, as well as on $\Gamma[f_0(980) \to \pi\pi]$.

by appropriately matching our theoretical prediction to the experimental data. A consequence of introducing the nonet (5) is that the scalar couplings are now related to each other by the underlying chiral symmetry, i.e. $\gamma_{spp} = \gamma_{spp}(A, B, C, D, \theta_s, \theta_p)$, where θ_p is the pseudoscalar mixing angle for which we choose the value 37°.

We numerically search through both ranges of θ_s and fit our prediction for πK scattering amplitude to the experimental data. This determines A and B in the interaction Lagrangian. We find that the χ^2 of fit improves as we lower m_κ. This is shown in Fig. 2, together with the m_κ dependence of the parameters A and B in the interaction Lagrangian, and the total decay width of κ. Although the χ^2 fit improves for lower values of m_κ, other experimental data further restricts the acceptable range of m_κ. We take into account limits from $\pi\pi$ scattering amplitude [6,7] on the strong interaction couplings γ_{spp}, as well as the decay width $\Gamma[f_0(980) \to \pi\pi]$. The m_κ dependence of the resulting couplings are shown in

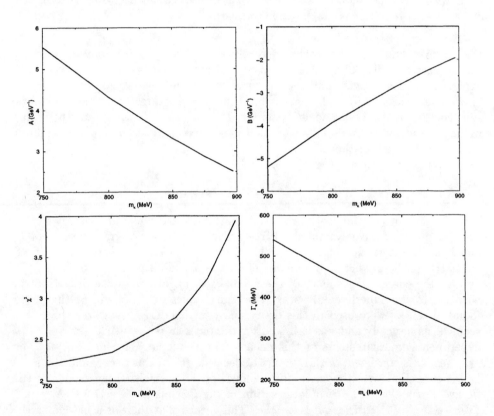

FIGURE 2. m_κ dependence of the fitting parameters A, B and Γ_κ, in a fit of the theoretical prediction of the πK scattering amplitude to the experimental data, together with the χ^2 of the fit.

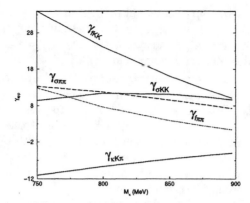

FIGURE 3. m_κ dependence of the hadronic couplings found in a fit of the theoretical prediction of the πK scattering amplitude to the experimental data.

Fig. 3. We find that the small angle solution is favored as it contains a small region (corresponding to 875 MeV $\leq m_\kappa \leq$ 897 MeV) that is consistent with these experimental constraints. This region, which is bounded by dashed lines in Fig. 1, is obviously close to $\theta_s = 0$. This is how our model indirectly probes the quark substructure of this scalar nonet – the fact that θ_s is small means that the scalar mixing in our model is closer to the dual ideal mixing and therefore a four-quark scenario is favored for this nonet.

IV $\eta' \to \eta\pi\pi$ DECAY

In the previous section we rewrote the scalar part of the Lagrangian in terms of a scalar nonet N. We evaluated all free parameters in the Lagrangian except C and D in the interaction piece in (7). These two parameters were probed in ref. [3] by matching the prediction of our model for the partial decay width of $\eta' \to \eta\pi\pi$, and for the energy dependence of the normalized magnitude of the decay matrix element, to the experimental data. The same values of A, B and θ_s that were found in ref. [2] were used in this decay analysis. The CD parameter space was scanned numerically and was searched for the physical regions that describe both experimental measurements of this decay. The result is shown in Fig. 4. The gray region is consistent with the partial decay width of this decay, and the solid line is consistent with the energy dependence of the normalized magnitude of the decay matrix element. Their intersection in the CD plane ($C \approx 7.3$ GeV^{-1} and $D \approx -1.7$ GeV^{-1}) *exists* and is *unique*. This means that there is a unique choice of free parameters of Lagrangian (7) that, in addition to $\pi\pi$ and πK scattering amplitudes, describes the $\eta' \to \eta\pi\pi$ decay. The energy dependence of this decay is plotted in Fig. 5.

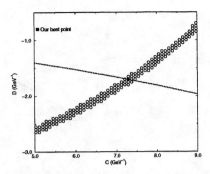

FIGURE 4. Regions in CD plane consistent with two different experimental measurements on η' decay. Regions consistent with $\Gamma^{exp.}[\eta' \to \eta\pi\pi]$ are represented by circles. The solid line represents points consistent with the energy dependence of the normalized magnitude of the decay matrix element.

As a by-product we compute, with the same extracted C and D, the partial decay width of $a_0(980) \to \pi\eta$ to be approximately 65 MeV. This, together with the $\Gamma[a_0(980) \to K\bar{K}] \approx 5 MeV$ found in [2], provide an estimate of the total decay width of $a_0(980)$ around 70 MeV. This is in a very close agreement with a recent experimental analysis of $a_0(980)$ in ref. [9].

V SUMMARY AND DISCUSSION

In this talk we reviewed the light scalar mesons in the non-linear chiral Lagrangian framework of references [1-3]. We saw that in this approach there is a need for the $\sigma(560)$ and the $\kappa(900)$ in order to be able to describe the experimental data on the $\pi\pi$ and πK scattering. We then constructed a light scalar nonet below 1 GeV consisting of $\sigma(560)$, $\kappa(900)$, $f_0(980)$, and $a_0(980)$, and rewrote the Lagrangian in terms of this nonet by introducing eight new free parameters. We showed that with these parameters we can describe many experimental facts including the scalar mass spectrum, their interactions with pseudoscalars in $\pi\pi$ and πK scattering, the η' decay, and the decay width of $f_0(980)$. We could predict the total decay width of $a_0(980) \approx 70$ MeV in a very close agreement with a recent experiment [9]. We discussed that although the chiral Lagrangian model presented here is entirely formulated in terms of the meson fields and in principle does not know anything about the underlying quark substructure, in practice, the knowledge of the mixing angle indirectly probes the quark substructure of these scalars. We saw, through a careful numerical analysis, that the acceptable range of the mixing angle is such that it suggests the quark substructure of these scalars is closer to

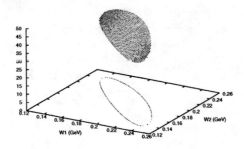

FIGURE 5. Energy dependence of the magnitude of the $\eta' \to \eta\pi\pi$ decay matrix element. w_1 and w_2 are the total energy of the final state pions, and are bounded within the ellipse-like region in the $\omega_1 \omega_2$ plane.

a four quark picture. This is in agreement with a recent theoretical investigation [10].

ACKNOWLEDGMENTS

This talk was based on works done in collaboration with D. Black, F. Sannino and J. Schechter [1–3], and has been supported in part by DE-FG-02-92ER-40704. We also thank the organizers of the MRST'99 for a very interesting conference.

REFERENCES

1. Black, D., Fariborz, A.H., Sannino, F., and Schechter, J., *Phys. Rev. D* **58**, 054012-1-11 (1998).
2. Black, D., Fariborz, A.H., Sannino, F., and Schechter, J., *Phys. Rev. D* **59**, 074026-1-12 (1999).
3. Fariborz, A.H., and Schechter, J., *Phys. Rev. D*, **60**, 034002-1-11 (1999).
4. Review of Particle Physics, *Euro. Phys. J. C* **3** (1999).
5. Asner, D.M., et al, CLEO Collaboration, hep-ex/9902022.
6. Sannino, F., and Schechter, J., *Phys. Rev. D* **52**, 96-107 (1995).
7. Harada, M., Sannino, F., and Schechter, J., *Phys. Rev. D* **54**, 1991-2005 (1996); *Phys. Rev. Lett.* **78**, 1603 (1997).
8. Jaffe, R.L., *Phys. Rev. D* **15**, 267-289 (1977).
9. Teige, S., et al, *Phys. Rev. D* **54**, 012001-1-12 (1999).
10. Achasov, N.N., *Phys. Usp.* **41**, 1149-1153 (1998); Achasov, N.N., and Shestakov, G.N., hep-ph/9904254.

The Theta Term in QCD Sum Rules and Electric Dipole Moments of Hadrons[1]

Maxim Pospelov and Adam Ritz

Theoretical Physics Institute, School of Physics and Astronomy
University of Minnesota, 116 Church St., Minneapolis, MN 55455, USA

Abstract. Using the QCD sum rule approach, we calculate the electric dipole moments of the ρ^+ meson and the neutron induced by a vacuum θ-angle to approximately 40-50% precision. In the latter case, we find $d_n = 1.2 \times 10^{-16} \bar{\theta} e \cdot cm$, which combined with the new experimental bound, translates into the limit $|\bar{\theta}| < 6 \times 10^{-10}$.

I INTRODUCTION

The impressive experimental limits on the electric dipole moments (EDMs) of neutrons and heavy atoms put very strong constraints on possible flavor-conserving CP-violation around the electroweak scale [1]. This precision means that EDMs can in principle probe a high energy scale by limiting the coefficients of operators with dimension≥ 4, such as $G\tilde{G}$, $GG\tilde{G}$, and $\bar{q}G\sigma\gamma_5 q$ etc. However, in practice, while these operators can be perturbatively evolved down to a scale of order 1 GeV, the ultimate connection between high energy parameters and low energy EDM observables necessarily involves non-perturbative physics.

The θ-term, $\theta G\tilde{G}$, is an important example of such CP violating operators, and among CP violating observables the EDM of the neutron ($d_n(\theta)$) is known to be the most sensitive to its value [3,4]. The calculation of $d_n(\theta)$ is, however, a long standing problem [5–7], with the chiral loop approach of Ref. [6] currently providing the most reliable results. This technique, however, relies of the dominance of contributions proportional to $\ln m_\pi$ near the chiral limit, and possible corrections may be large in a more realistic situation.

There are important incentives for refining the calculation of $d_n(\theta)$. If θ is zero at tree level, either due for example to exact P or CP symmetries [8], or via the axion mechanism, one still induces corrections to θ through either radiative effects or higher dimensional CP-odd operators [9]. In this case, the calculation of $d_n(\theta)$ is part of a more generic calculation of d_n, needed, for example, in the MSSM for extraction of limits on CP-violating SUSY phases.

---

[1] Talk presented by M. Pospelov

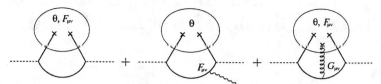

FIGURE 1. Contributions to the correlator at leading order in $F_{\mu\nu}$.

In this report, we describe a recent [14,18] application of the QCD sum rule method [2] to obtain an estimate for $d_n(\theta)$ beyond chiral perturbation theory. The sum rules formalism appears suited to this task as it has been used previously to calculate certain baryonic electromagnetic form factors [10,11], and also the neutron EDM induced by a CP-odd color electric dipole moment of quarks [12,13]. Before turning our attention to the neutron, we first discuss in Section II the application of this approach to ρ-meson, as a useful testing ground for the technique [14]. In Section III we then turn to the neutron, obtaining a result for $d_n(\theta)$ which is actually quite close to the chiral-loop estimate. We refer the reader to [14,18] for full details of these calculations.

II CALCULATION OF THE ρ–EDM

Since it is a spin 1 particle, the ρ-meson can possess on-shell two CP-odd electromagnetic form factors, the electric dipole and magnetic quadrupole moments. We shall concentrate here on the EDM of $\rho^{+(-)}$ as the CP-violating form factors of ρ^0, induced by the theta term, vanish as a consequence of C-symmetry.

In order to calculate the ρ^+ EDM within the sum rule approach, we need to consider the correlator of currents with ρ^+ quantum numbers, in a background with nonzero θ and an electromagnetic field $F_{\mu\nu}$,

$$\Pi_{\mu\nu}(Q^2) = i\int d^4x\, e^{iq\cdot x}\langle 0|T\{j_\mu^+(x)j_\nu^-(0)\}|0\rangle_{\theta,F}, \qquad (1)$$

where $j_\mu^+ = \bar{u}\gamma_\mu d$, and we denote $Q^2 = -q^2$, with q the current momentum. To simplify matters, we consider the $m_u = m_d$ limit for quark masses, although we keep the explicit mass dependence below, and will comment on the unequal mass case at the end of the section.

The relevant diagrams to be evaluated are shown in Fig. 1. We work with a constant background field $F_{\mu\nu}$ and a fixed point gauge for the gluon field, and extract the appropriate Lorentz structure ($\tilde{F}_{\mu\nu}$) by projecting onto vacuum condensates. The dependence on $F_{\mu\nu}$ is determined via introduction of the magnetic susceptibilities χ, ξ and κ [10]. For example, $\langle 0|\bar{q}\sigma_{\mu\nu}q|0\rangle_F = \chi e_q F_{\mu\nu}\langle 0|\bar{q}q|0\rangle$ (see [10] for the definitions of ξ and κ).

The leading θ–dependence of these matrix elements is extracted via use of the anomalous Ward identity (see e.g. [16]). We recall here the resulting expression for a generic structure $m_q\langle\bar{q}\Gamma q\rangle_{\theta_G}$,

$$m_q \langle 0|\bar{q}\Gamma q|0\rangle_{\theta_G} = im_*\theta_G \langle 0|\bar{q}\Gamma\gamma_5 q|0\rangle + O(m_q^2), \tag{2}$$

where $m_* = m_u m_d/(m_u + m_d)$ for two flavours, and the ability to neglect the $O(m_q^2)$ corrections follows since $m_\eta \gg m_\pi$. The overall factor has the form $(1 - m_\pi^2/m_\eta^2)$ [16] which vanishes when $U(1)$-symmetry is restored ($m_\eta \to m_\pi$) [14,18]. We could also obtain this result of course by rotating θ to a γ_5-mass term.

Using these techniques, and performing the rather lengthy computations, our final expression for the OPE side of sum rule is,

$$\Pi_{\mu\nu}^+ = m_*\theta(e_u - e_d)\langle 0|\bar{q}q|0\rangle \left[\frac{\tilde{F}_{\mu\nu}}{q^2}\left(-\chi - \frac{1}{q^2}\left(1 + \kappa - \frac{\xi}{4}\right)\right)\right]. \tag{3}$$

On the phenomenological side, we parametrize in the usual manner the double and single pole contributions, the latter representing transitions between the neutron and excited states. After a Borel transform (see e.g. [19]), the sum rule has the form,

$$\lambda^2 f_1 + AM^2 = \frac{1}{2}m_*\theta(e_u - e_d)M^4 e^{m_\rho^2/M^2}\langle 0|\bar{q}q|0\rangle\left(\frac{\chi}{M^2} - \frac{1}{M^4}\left(1 + \kappa - \frac{\xi}{4}\right)\right). \tag{4}$$

where f_1 determines the EDM, λ is the coupling of the current to ρ^+, and A parametrizes the single pole terms. The physical EDM parameter d_ρ is given by

$$d_\rho = \frac{f_1}{m_\rho} \equiv \tilde{d}\frac{m_*}{m_\rho}\theta(e_u - e_d), \tag{5}$$

where $\tilde{d}$ has also been introduced for convenience.

For numerical calculation we make use of the following parameter values: For the quark condensate, we have $\langle 0|\bar{q}q|0\rangle = -(0.225 \text{ GeV})^3$, while for the condensate susceptibilities, we have: $\chi = -5.7 \pm 0.6 \text{ GeV}^{-2}$ [20], $\kappa = -0.34 \pm 0.1$ [13], and $\xi = -0.74 \pm 0.2$ [13].

The parameter A may be removed by differentiation [14], and the coupling λ may be determined by using the corresponding CP-even sum rule [19,14]. Consequently, we find (with $s_0 \sim (1.7\text{GeV})^2$, and $\langle \mathcal{O}_4 \rangle \equiv \langle 0|\frac{\alpha_s}{\pi}G^2|0\rangle + 24\langle 0|m_q\bar{q}q|0\rangle$),

$$\tilde{d} = 2\pi^2 \frac{\langle 0|\bar{q}q|0\rangle(s_0 + M^2 - m_\rho^2)}{(s_0(1 + \alpha_s/\pi) + \pi^2(s_0 + 2M^2)\langle \mathcal{O}_4\rangle/M^4)}\left(\frac{\chi}{M^2} - \frac{1}{M^4}\left(1 + \kappa - \frac{\xi}{4}\right)\right). \tag{6}$$

Various contributions to this sum rule are shown in Fig. 2. $\tilde{d}_1$ represents an estimate in which all higher order corrections are ignored, $M^2 = m_\rho^2$, and λ is fixed numerically. Removing one level of approximation, we return to (6) and set all the $O(1/M^4)$ corrections zero, obtaining $\tilde{d}_2$. The leading correction ($\tilde{d}_3$) at $O(1/M^4)$ may be isolated by setting $\kappa = \xi = 0$ in (6). The presence of the $1/M^4$ term induces a transition region in the M^2 dependence. Finally, we can obtain an estimate

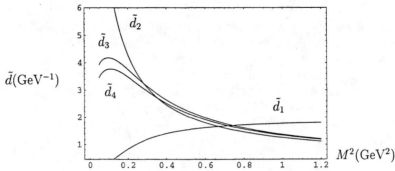

FIGURE 2. The ρ^+ EDM parameter $\tilde{d}$ as a function of M^2 according to various components of the sum rule (6).

of the corrections associated with κ and ξ, by plotting the full expression in (6), which is displayed in Fig. 2 as $\tilde{d}_4$. The corrections are encouragingly quite small. Extracting a numerical estimate for $\tilde{d}$, we find the result

$$d_\rho = (2.6 \pm 0.8) e\theta \frac{m_*}{(1\text{GeV})^2}, \qquad (7)$$

for the EDM of ρ^+, where $e = e_u - e_d$ is the positron charge.

To conclude this section, we shall make some comments on the quark mass dependence if $m_u \neq m_d$. If one starts with a basis in which θ is rotated to lie in front of the quark singlet combination $\bar{q}i\gamma_5 q$, then the result will automatically appear with the correct m_* mass dependence. If instead, one leaves θ in front of $G\tilde{G}$, then to restore the correct mass dependence to d_ρ, it is necessary to account for the fact that the current itself is not chirally invariant. Thus there are additional sources of mixing with axial-vector correlators (through the anomaly). Once this mixing is taken into account one does indeed recover a result proportional to m_* as expected (see [14] for details).

III CALCULATION OF THE NEUTRON EDM

We now turn our attention to the primary goal of obtaining the neutron EDM. The approach we shall use is similar to that successfully implemented for ρ^+, so we focus here on certain additional subtleties which make the problem more involved. We start as usual with the correlator of currents $\eta_n(x)$ with quantum numbers of the neutron,

$$\Pi(Q^2) = i \int d^4x e^{iq\cdot x} \langle 0|T\{\eta_n(x)\bar{\eta}_n(0)\}|0\rangle_{\theta,F}. \qquad (8)$$

The first point to note is that if CP-symmetry is broken by a generic quark-gluon CP-violating source (θ–term in our case), the coupling between the physical state

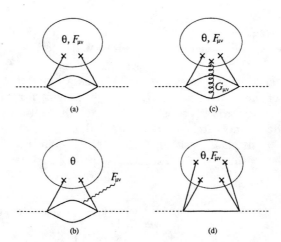

FIGURE 3. Various contributions to the CP-odd structure $\{\tilde{F}\sigma, \not{q}\}$. (a) is the leading order contribution while (b) and (c) contribute at subleading order.

(neutron) described by a spinor v and the current η_n acquires an additional phase factor $\langle 0|\eta_n|N\rangle = \lambda U_\alpha v$, where $U_\alpha = e^{i\alpha\gamma_5/2}$. This unphysical phase α can unfortunately mix electric (d) and magnetic (μ) dipole moment structures and complicate the extraction of d from the sum rule. A consideration of all the tensor structures appearing in the double pole phenomenological contribution indicates that there is only one, $\{\tilde{F}\sigma, \not{q}\}$, which we shall therefore study here, for which d is isolated independent of the phase α.

The interpolating current η_n is conveniently parametrized in the form,

$$\eta_n = j_1 + \beta j_2, \qquad (9)$$

where the two contributions are given by

$$j_1 = 2\epsilon_{abc}(d_a^T C\gamma_5 u_b)d_c \qquad j_2 = 2\epsilon_{abc}(d_a^T C u_b)\gamma_5 d_c. \qquad (10)$$

j_2 vanishes in the nonrelativistic limit, and lattice simulations have shown that j_1 indeed provides the dominant projection onto the neutron (see e.g. [17]). Within the sum rules formalism, one also has the imperative of suppressing the contribution of excited states and higher dimensional operators in the OPE, and thus its convenient to choose β to this end. Ioffe has shown that $\beta = -1$ is an apparently optimal choice for the mass sum rule. However, it is clear that this optimization may differ for different physical observables. We shall therefore keep β arbitrary, and optimize once we have knowledge of the structure of the sum rule.

We now proceed to study the OPE associated with (8). The relevant diagrams we need to consider are shown in Fig. 3 ((a), (b) and (c)). Diagrams of the form (d), although suffering no loop factor suppression, are nonetheless suppressed due to combinatorial factors and the small numerical size of $(\langle\bar{q}q\rangle)^2$ (see [18]).

In parametrizing θ, we shall now take a general initial condition in which a chiral rotation has been used to generate a γ_5-mass, with coefficient θ_q, while we denote the coefficient of $G\tilde{G}$ by θ_G. The physical parameter is $\bar{\theta} = \theta_q + \theta_G$, but we shall keep the general form and calculate the OPE as a function of *both* phases. The independence of the final answer of $\theta_q - \theta_G$ will provide a nontrivial check on the consistency of our approach. We shall find that this requires the consideration of mixing with additional currents, a point we shall come to shortly.

The actual calculations, although considerably more involved, follow the approach discussed earlier for ρ^+. The final result [18] for the OPE structure takes the form

$$\Pi(Q^2) = -\frac{\bar{\theta} m_*}{64\pi^2}\langle\bar{q}q\rangle\{\tilde{F}\sigma, \slashed{q}\}\left[\chi(\beta+1)^2(4e_d - e_u)\ln\frac{\Lambda^2}{Q^2}\right.$$
$$-4(\beta-1)^2 e_d\left(1 + \frac{1}{4}(2\kappa+\xi)\right)\left(\ln\frac{Q^2}{\mu_{IR}^2} - 1\right)\frac{1}{Q^2}$$
$$\left.-\frac{\xi}{2}\left((4\beta^2 - 4\beta + 2)e_d + (3\beta^2 + 2\beta + 1)e_u\right)\frac{1}{Q^2}\cdots\right], \quad (11)$$

where μ_{IR} is an infrared cutoff.

At different stages of the calculation we observe the appearance of the unphysical phase $\theta_G - \theta_q$ which is closely related with the non-invariance of the currents j_1 and j_2 under chiral transformations. This invariance is, in fact, restored when we consider the mixing of j_1 and j_2 with another set of currents, $i_1 = 2\epsilon_{abc}(d_a^T C u_b) d_c$ and $i_2 = 2\epsilon_{abc}(d_a^T \gamma_5 C u_b)\gamma_5 d_c$. The mixing between these two sets is explicitly proportional to $\theta_G - \theta_q$, and when properly diagonalized on the sum rule for $\slashed{q}$, the linear combination of these two sets of currents provides additional contributions proportional to $\theta_G - \theta_q$, exactly cancelling the unphysical piece of $\Pi(Q^2)$, and leaving us with the expected $\bar{\theta}$ dependence.

Inspection of the expression (11) indicates that a natural choice for the current is $\beta = 1$ as it cancels the ambiguous subleading infrared logarithm. We note that the effect of this choice mimics that of the standard choice $\beta = -1$ (Ioffe current) for CP-even sum rules. Although we shall use $\beta = 1$, we point out that our results are not highly sensitive to β; use of the "lattice" current, for example, with $\beta = 0$ will also produce a numerically similar result.

On the phenomenological side of the sum rule we once again parametrize the double and single pole contributions, as discussed earlier for ρ^+. After a Borel transform, we find

$$\lambda^2 m_n d_n + AM^2 = -\frac{M^4}{64\pi^2}\bar{\theta}m_*\langle\bar{q}q\rangle e^{m_n^2/M^2}$$
$$\times \left[4\chi(4e_d - e_u) - \frac{1}{2M^2}\xi(4e_d + 8e_u)\right] \quad (12)$$

where d_n is the neutron EDM and $\lambda = \lambda_1 + \beta\lambda_2$.

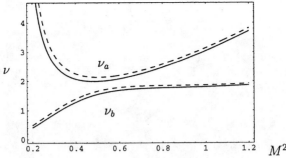

FIGURE 4. The neutron EDM function $\nu(M^2(\text{GeV}^2))$ is plotted according to the sum rules (a) and (b). The dashed line shows the contribution from the leading order term only.

The coupling λ present in (12) may be obtained from the well known sum rules for the tensor structures $\mathbf{1}$ and $\not{q}$ in the CP even sector (see e.g. [17] for a recent review). We shall construct two sum rules in this way.

- **(a)** Firstly, we extract a numerical value for λ via a direct analysis of the CP even sum rules. This analysis has been discussed before and will not be reproduced here (see e.g. [17]). One obtains $(2\pi)^4 \lambda \sim 1.05 \pm 0.1$.

- **(b)** As an alternative, we extract λ explicitly as a function of β from the CP-even sum rule for $\not{q}$, and substitute the result into (12) choosing $\beta = 1$.

It is convenient to define an additional function $\nu(M^2)$,

$$\nu(M^2) \equiv \frac{1}{\bar{\theta} m_*} \left(d_n + \frac{AM^2}{\lambda^2 m_n} \right), \quad (13)$$

which is then determined by the right hand side of (12) where we assume that A is independent of M^2.

The two sum rules described above for ν_a and ν_b are plotted in Fig. 4. One observes that both sum rules have extrema consistent to $\sim 20\%$, suggesting that our procedure for fixing the parameter β is appropriate. Furthermore, the differing behaviour away from the extrema implies that for consistency we must assume A to be small. Extracting a numerical estimate for d_n from Fig. 4, and determining an approximate error, we find the neutron EDM

$$d_n = \frac{(1.0 \pm 0.3)}{(700\text{MeV})^2} \bar{\theta} m_* = \frac{(90 \pm 30) f_\pi^2 m_\pi^2 m_u m_d}{(m_u + m_d)^2} \bar{\theta}, \quad (14)$$

for which the dominant contribution naturally arises from χ.

Comparison with the result of Ref. [6] indicates rather good agreement in magnitude, due essentially to the low effective mass scale $M \sim 700\text{MeV}$.

IV DISCUSSION

Combining our result for the neutron EDM with the recently improved experimental bound on d_n [4] we derive, allowing a 40-50% uncertainty, the limit on theta:

$$|\bar{\theta}| < 6 \times 10^{-10}, \tag{15}$$

which is quite close to previous bounds.

In conclusion, we have presented QCD sum rules calculations of the θ-induced EDMs for ρ^+ and the neutron. The results are explicitly tied to a set of vacuum correlators which are non-vanishing only in the absence of a U(1) "Goldstone boson". For the neutron, the use of QCD sum rules in the chirally invariant channel allowed us to unambiguously extract $d_n(\theta)$, and independence of the answer from any particular representation of the theta term was checked explicitly.

Acknowledgments We would like to thank M. Shifman and A. Vainshtein for many valuable discussions and comments.

REFERENCES

1. I. B. Khriplovich and S. K. Lamoreaux, *"CP Violation Without Strangeness"*, Springer, 1997.
2. M.A. Shifman, A.I. Vainshtein and V.I. Zakharov, Nucl. Phys. **B147** 385; 448 (1979).
3. K.F. Smith *et al.*, Phys. Lett. **B234** 191 (1990); I.S. Altarev *et al.*, Phys. Lett. **B276** 242 (1992).
4. P.G. Harris *et al.*, Phys. Rev. Lett. **82** 904 (1999).
5. V. Baluni, Phys. Rev. **D19** 2227 (1979).
6. R.J. Crewther *et al.* Phys. Lett. **B88** 123 (1979); **91** 487(E) (1980).
7. S. Aoki and T. Hatsuda, Phys. Rev. **D45** 2427 (1992).
8. see e.g. R.N. Mohapatra, Phys. Atom. Nucl. **61**, 963 (1998) and references therein.
9. I. Bigi and N.G. Uraltsev, Sov. Phys. JETP **100** 198 (1991); M. Pospelov, Phys. Rev. **D58** 097703 (1998).
10. B.L. Ioffe and A.V. Smilga, Nucl. Phys. **B232** 109 (1984).
11. I.I. Balitsky and A.V. Yung, Phys. Lett. **B129** 328 (1983).
12. V.M. Khatsimovsky, I.B. Khriplovich, and A.S. Yelkhovsky, Ann. Phys. **186** 1 (1988).
13. I.I. Kogan and D. Wyler, Phys. Lett. **B274** 100 (1992).
14. M. Pospelov and A. Ritz, hep-ph/9903553.
15. R.J. Crewther, Phys. Lett. **70B** 349 (1977).
16. M.A. Shifman, A.I. Vainshtein and V.I. Zakharov, Nucl. Phys. **B166** 493 (1980).
17. D. B. Leinweber, Ann. Phys. **254**, 328 (1997).
18. M. Pospelov and A. Ritz, hep-ph/9904483; TPI-MINN-99/34 (1999), to appear.
19. L.J. Reinders, H. Rubinstein, and S. Yazaki, Phys. Rept. **127** 1 (1985).
20. V.M. Belyaev and Ya.I. Kogan, Sov. J. Nucl. Phys. **40** 659 (1984).

The Supersymmetric Spectrum of Gauge-Mediated Supersymmetry Breaking of $SO(10)$

M. Frank[a], H. Hamidian[b], K. Puolamäki[c]

[a]*Department of Physics, Concordia University, Montreal, Quebec, Canada H3G 1M8*

[b]*Department of Physics, University of Illinois at Chicago, Chicago, Illinois, 60607-7059*

[c]*Helsinki Institute of Physics, P.O.Box 9, FIN-00014 University of Helsinki, Finland*

Abstract. We investigate gauge-mediated supersymmetry breaking of $SO(10)$ into either $SU(2)_L \times SU(2)_R \times U(1)_{B-L}$ or $SU(2)_L \times U(1)_{I_{3R}} \times U(1)_{B-L}$. We impose perturbativity of the gauge couplings and unification at the GUT scale severely restrict the messenger sector and obtain predictions for the sparticle masses.

Supersymmetry (SUSY) is currently believed to lead to the most attractive scenario of physics beyond the standard model (SM). However, the low-energy spectrum of fermions and bosons does not exhibit this symmetry, so SUSY must somehow be broken. A popular realization of this breaking can be accommodated through gravity, and this mechanism has been explored extensively [1]. In gravity-mediated breaking, one introduces a hidden sector (gravity) which breaks supersymmetry through interactions with the visible sector. The gravitational interactions operate at M_{Planck} and they trigger the breaking through soft-breaking terms. The drawbacks of these type of breaking is that one must deal with flavour-changing-neutral-currents (FCNC) associated with soft quark and slepton masses, and of course, one must deal with the divergencies associated with gravity.

Another possibility is the gauge-mediated supersymmetry breaking (GMSB) [2] in which the mechanism used to communicate SUSY breaking is provided by keeping the original theory renormalizable, but with a low-energy description in terms of an effective Lagrangian with non-renormalizable terms. This generates soft terms at some "messenger" scale below the Planck scale and breaks flavor symmetry only through Yukawa couplings.

To carry out the GMSB program, one starts with an observable sector which contains the usual matter and gauge fields and their supersymmetric partners, while leaving the hidden sector unspecified. One then introduces a *messenger* sector, formed by the new superfields, whose coupling with the goldstino superfield generates a supersymmetric mass M for the messenger fields, and leads to mass splittings of order F (where we denote by F the vacuum expectation value of the auxiliary component of the superfield), and thus $\sqrt{F}$ is identified with the scale of SUSY breaking in the messenger sector [3]. Apart from the requirement that the messenger fields transform under the SM gauge group, the messenger sector is unknown and is the main source of model-dependence in GMSB theories.

Explicit model building in theories which take advantage of the GMSB mechanism has been explored in a number of papers by taking $SU(5)$ to be the unifying gauge group [4]. However $SU(5)$ has shortcomings related to problems such as the nucleon decay rates and neutrino masses, as well as R-parity breaking. Here we present a study of the GMSB mechanism and its phenomenological consequences in an $SO(10)$ supersymmetric grand unified theory (GUTs).

The $SO(10)$ gauge group is a natural candidate for supersymmetric unification [5] since all the quarks and leptons of a single generation are the components of a single spinor representation and provides non-zero neutrino masses through the see-saw mechanism.

The symmetry breaking chain that we study here is as follows. We assume that the $SO(10)$ gauge group is broken to an intermediate left-right symmetry group G_{LR} at scale $M_{\dot{G}}$. M_G should be no less than 10^{16} GeV to have stable nuclei and it should be below the Planck scale 10^{19} GeV. The intermediate symmetry group G_{LR} is then broken down to the MSSM at scale M_R. For simplicity, and to maintain only a minimal number of scales in the theory, we will assume that there are no further symmetry breaking scales between M_R and M_G.

As possible left-right symmetry groups we will consider $G_{LR}^I = SU(3)_C \times SU(2)_L \times U(1)_{I_{3R}} \times U(1)_{B-L}$ and $G_{LR}^{II} = SU(3)_C \times SU(2)_L \times SU(2)_R \times U(1)_{B-L}$.

The model based on the gauge group $SU(2)_L \times U(1)_{I_{3R}} \times U(1)_{B-L}$ is phenomenologically interesting because it contains all the usual matter multiplets *plus* the right-handed neutrinos and forbids lepton- and baryon-number violating terms in the superpotential. The superpotential for the matter sector of the theory is:

$$W = h_u Q H_u u^c + h_d Q H_d d^c + h_e L H_d e^c + h_\nu L H_u \nu^c$$
$$+ \mu H_u H_d + f \delta \nu^c \nu^c + M_R \delta \bar{\delta} + W_m, \qquad (1)$$

where W_m denotes the messenger sector superpotential. The lepton and the quark sectors in this theory consists of the doublets $Q(2, 0, 1/3)$, $L(2, 0, -1)$, and the singlets $u^c(1, -1/2, -1/3)$, $d^c(1, 1/2, -1/3)$, $e^c(1, 1/2, 1)$, $\nu^c(1, -1/3, 1)$, and the corresponding squarks and sleptons. The two Higgs doublets (and their superpartners) in this model are the same as in the MSSM: $H_u(2, 1/2, 0)$ and $H_d(2, -1/2, 0)$. In addition to these, the model contains two Higgs triplets $\delta(1, 1, -2)$ and $\bar{\delta}(1, -1, 2)$ which break the $U(1)_{I_{3R}} \times U(1)_{B-L}$ symmetry down to the $U(1)_Y$ of the SM, together with their superpartners. The gauge sector of the the model contains the bosons $W(3, 0, 0)$, $B(1, 0, 0)$ and $V(1, 0, 0)$, and the corresponding gauginos.

The left-right supersymmetric model based on the gauge group $SU(2)_L \times SU(2)_R \times U(1)_{B-L}$ has been studied intensively over the years [6]. In addition to the attractive features of Model I, this model offers possible solutions to both the strong and weak CP problems [7], while at the same time preserving R-parity. The superpotential for the matter sector of this theory is:

$$W = \mathbf{h}_q^{(i)} Q_L^T \tau_2 \Phi_i \tau_2 Q_R + \mathbf{h}_l^{(i)} L_L^T \tau_2 \Phi_i \tau_2 L_R + i(\mathbf{h}_{LR} L_L^T \tau_2 \delta_L L_L + \mathbf{h}_{LR} L_R^T \tau_2 \Delta_R L_R)$$
$$+ M_{LR}\left[Tr(\Delta_L \bar{\delta}_L + Tr(\Delta_R \bar{\delta}_R)\right] + \mu_{ij} Tr(\tau_2 \Phi_i^T \tau_2 \Phi_j) + W_m, \tag{2}$$

where, as before, W_m denotes the messenger sector superpotential. The particle content in this theory is as follows. The lepton and the quark sectors consist of the doublets $Q_L(2, 1, 1/3)$, $L_L(2, 1, -1)$, $Q_R(1, 2, -1/3)$, $L_R(1, 2, 1)$, and the corresponding superpartners. This model contains bi-doublet Higgs fields $\Phi_u(2, 2, 0)$ and $\Phi_d(2, 2, 0)$, the triplet Higgs fields $\Delta_L(3, 1, -2)$ and $\Delta_R(1, 3, -2)$ which break the left-right model to the SM, as well as $\delta_L(3, 1, 2)$, $\delta_R(0, 3, 2)$ (which are required for the cancellation of anomalies in the fermionic sector), and their superpartners. The gauge sector consists of the bosons $W_L(3, 1, 0), W_R(1, 3, 0), B(1, 0, 0)$, and $V(1, 1, 0)$ and their associated superpartners.

The simplest possibility to break SUSY which avoids the proliferation of scales is to assume that the SUSY breaking scale and the left-right gauge symmetry breaking scale could be somehow related. To realize this we will further assume that the SUSY breaking scale Λ_{SUSY} is the same as the left-right symmetry breaking scale M_R. This choice simultaneously connects the scale of the gauge symmetry breaking to the scale of SUSY breaking and fulfills the requirement that the breaking of the electroweak symmetry remain radiative. In order for the sparticle masses to be around 1 TeV, the SUSY breaking scale must be $\Lambda_{\text{SUSY}} \sim M_R \sim 100$ TeV.

The simplest messenger sector consists of N_f flavors of chiral superfields Φ_i and $\bar{\Phi}_i$, $(i = 1, \cdots, N_f)$ which transform in the $\mathbf{r} + \bar{\mathbf{r}}$ representation of the gauge group. In order to preserve gauge coupling constant unification, one usually requires that the messengers form complete GUT multiplets. The minimal set of messenger fields can be relaxed and augmented further if one requires a theory sufficiently rich to be viable. On the other hand, the number of messenger fields is restricted by the requirement that the MSSM couplings stay perturbative up to the GUT scale and that gauge couplings unify to a single coupling at M_G. The possible messenger fields in Model I are: $Q_8 = (8, 1, 0, 0), L_3 = (1, 3, 0, 0), \Delta + \bar{\Delta} = (1, 3, 0, -2) + conj.$, $\Delta^c + \bar{\Delta^c} = (1, 1, -1, 2) + conj.$, $H + \bar{H} = (1, 2, \frac{1}{2}, 0) + conj.$, $Q + \bar{Q} = (3, 2, 0, \frac{1}{3}) + conj.$, $U^c + \bar{U^c} = (\bar{3}, 1, -\frac{1}{2}, -\frac{1}{3}) + conj.$, $D^c + \bar{D^c} = (\bar{3}, 1, \frac{1}{2}, -\frac{1}{3}) + conj.$, $L + \bar{L} = (1, 2, 0, -1) + conj.$, $e^c + \bar{e^c} = (1, 1, \frac{1}{2}, 1) + conj.$, $\nu^c + \bar{\nu^c} = (1, 1, -\frac{1}{2}, 1) + conj.$, where their $SU(3)_C \times SU(2)_L \times U(1)_{I_{3R}} \times U(1)_{B-L}$ quantum numbers are specified in brackets.

Similarly, the possible messenger fields in Model II consist of $Q_8 = (8, 1, 1, 0), Q_3 = (1, 3, 1, 0), Q_3^c = (1, 1, 3, 0), \phi = (1, 2, 2, 0), Q + \bar{Q} = (3, 2, 1, \frac{1}{3}) + conj.$, $Q^c + \bar{Q^c} = (3, 1, 2, -\frac{1}{3}) + conj.$, $L + \bar{L} = (1, 2, 1, -1) + conj.$, $L^c + \bar{L^c} =$

$(1, 1, 2, 1) + conj.$, $\Delta + \overline{\Delta} = (1, 3, 1, -2) + conj.$, $\Delta^c + \overline{\Delta^c} = (1, 1, 3, 2) + conj.$, whre their $SU(3)_C \times SU(2)_L \times SU(2)_R \times U(1)_{B-L}$ quantum numbers are specified in brackets.

In the following section we shall restrict the messenger sectors by requiring (i) the perturbativity of the gauge couplings ($\alpha_k < 1$) up to the Planck scale and (ii) unification at the GUT scale M_G.

Denoting by β_L, β_R, β_V and β_C the beta-functions corresponding to the $SU(2)_L$, $SU(2)_R$ (or $U(1)_{I_{3R}}$), $U(1)_V$, and $SU(3)_C$ gauge groups respectively, the one-loop renormalization group equations at the M_R scale are as follows:

$$\alpha_i^{-1}(M_R) = \alpha_G^{-1} + \beta_L(t_G - t_i) \tag{3}$$

where $i = L, R, V, C$ and we defined:

$$t_R = \frac{1}{2\pi} \ln \frac{M_R}{M_Z} \quad \text{and} \quad t_G = \frac{1}{2\pi} \ln \frac{M_G}{M_Z}. \tag{4}$$

Using the one-loop matching conditions to the MSSM at the scale M_R, we obtain the following limits on the differences of LR β-functions:

$$3.1 < \beta_2^{LR} - \beta_3^{LR} < 4.1 \text{ and } 7.4 < \beta_1^{LR} - \beta_3^{LR} < 9.9. \tag{5}$$

The requirement of the perturbativity of the gauge couplings at the scale M_G and between M_G and M_P, leads to the following constraints on the left-right β-functions:

$$\beta_1^{LR} < 10.4, \ \beta_2^{LR} < 6.1 \text{ and } \beta_3^{LR} < 3.0. \tag{6}$$

These equations constrain the number of messenger fields, since each new chiral superfield contributes to the β-functions.

Using the β-functions for Models I and II [8], we find that there can be no consistent solutions in Model II. In Model I, on the other hand, there exist consistent solutions with the messenger multiplicities:

$$n_8 = n_3 = n_H + n_L = 1 \text{ and } n_{e^c} + n_{\nu^c} = 0, 1. \tag{7}$$

Studying these possibilities in more detail one sees that despite apparently having several choices, their consequences are remarkably alike for all the models. We have listed a sample of the spectrum obtained from these models in Table I (for $\Lambda_{SUSY} = 100\ TeV, \Lambda_M = 100\ \Lambda_{SUSY}$).

In every case studied the $SU(2)_L$ and the $SU(3)_C$ gauge couplings meet at $M_G = 2.0 \times 10^{16}$ GeV (which we take to be our GUT scale) for 10^5 GeV $< M_R < 10^7$ GeV.

Despite the fact that our restrictions allow six different solutions, we find that all of them are somewhat remarkably similar in predictions for the supersymmetric spectrum. The characteristic features of the obtained mass spectrum for the super-symmetric partners (and $H^\pm$) are: (i) Depending on the exact messenger content, the next-to-lightest supersymmetric particle (NLSP) can be either the lighter stau or the lightest neutralino, which is essentially a gaugino. The choice of $n_{e^c} = 1$ favours the neutralino as the NLSP, while solutions with $n_{e^c} = 0$ favour the stau as the NLSP. (ii) As expected, the lighter selectron is always heavier than the lighter stau. Also, the scalar leptons are always lighter than the squarks, which turn out to be very heavy in this model. (iii) The sneutrinos are always heavier than the lighter of the charged sleptons, unlike in supersymmetric models without GMSB. In fact $m_{\tilde{\nu}_{e,\tau}} \approx m_{\tilde{\tau}_2}$. The charged sleptons are lighter than the Higgs/Higgsinos. (iv) The bilinear Higgs coupling, the so-called μ parameter in the superpotential, lies in the $400 - 500$ GeV region and can be *either* positive *or* negative. $\mu \cong M_{\tilde{q}}/100$, meaning that $|\mu|$ is always at least 50 GeV. This makes the Higgsino at least twice as heavy as the wino, so that the light charginos are mostly gauginos. In general, the sign of μ does not seem to make much difference to the mass spectrum. However, we obtain sizable effects in the $b \to s\gamma$ decay width, since for positive μ the interference between the SM and chargino contributions is destructive, whereas for negative μ it is constructive. (v) The heavy spartner masses are quite accurately directly proportional to the scale $\Lambda_{\text{SUSY}} = F/S$. The heavy sleptons, neutralinos and charginos all have masses between 310 GeV and 470 GeV.

In summary, we have chosen to study supersymmetric $SO(10)$ GUTs with GMSB as viable SUSY GUT candidates, since these are free from the problems that plague $SU(5)$-based theories, with or without GMSB. By breaking $SO(10)$ to left-right symmetric gauge groups, and by taking advantage of the extreme predictive power of the GMSB mechanism, we calculate and discuss a number of important phenomenological results. The supersymmetric spectrum obtained is not unlike the one obtained in gauge-mediated models with a simpler breaking chain. This raises the attractive possibility that gauge-mediated breaking imposes similar features on a variety of GUT scenarios. It would certainly be interesting to study other GUT scenarios with GMSB to test what appear to be general features of this mechanism.

We would like to thank the organizers, Pat Kalyniak, Steve Godfrey and Basim Kamal for an enjoyable conference and wish Professor Sundaresan many more productive years in research.

REFERENCES

1. P. Nath, R. Arnowitt and A. H. Chamseddine, *Applied N=1 Supergravity* (World Scientific, Singapore, 1984).; H. P. Nilles, Phys.Rep.**110**(1984).
2. M. Dine, W. Fischler, and M. Srednicki, Nucl. Phys. **B189** (1981) 575; S. Dimopoulos and S. Raby, Nucl. Phys. **B192** (1981) 353; M. Dine and W. Fischler, Phys. Lett. **110B** (1982) 227; M. Dine and M. Srednicki, Nucl. Phys. **B202** (1982) 238; L. Alvarez-Gaumé, M. Claudson, and M. Wise, Nucl. Phys. **B207** (1982) 96; C. Nappi and B. Ovrut, Phys. Lett. **113B** (1982) 175.
3. M. Dine and A.E. Nelson, Phys. Rev. **D48** (1993) 1277; M. Dine, A.E. Nelson, and Y. Shirman, *ibid.* **D51** (1995) 1362; M. Dine, A.E. Nelson, Y. Nir, and Y. Shirman,

ibid. **D53** (1996) 2658. See also, G.F. Giudice and R. Rattazzi, hep-ph/9801271 for review and further references.

4. S. Dimopoulos, S. Thomas, and J. D. Wells, Phys. Rev. **D54** (1996) 3283; K. S. Babu, C. Kolda, and F. Wilczek, Phys. Rev. Lett. **77** (1996) 3070; J. A. Bagger, K. Matchev, D. M. Pierce, and R.-J. Zhang, Phys. Rev. **D55** (1997) 3188; H. Baer, M. Brhlik, C.-H. Chen, and X. Tata, Phys. Rev. **D55** (1997) 4463; N. G. Deshpande, B. Dutta, and S. Oh, Phys. Rev. **D56** (1997) 519.
5. K.S. Babu and S.M. Barr,Phys. Rev. **D48** (1993) 5354; Phys. Rev. **D50** (1994) 3529; Phys. Rev. **D51** (1995) 2463; Y. Nomura, T. Yanagida, Phys. Rev. **D59** (1999) 017303; Y. Nomura, T. Sugimoto, hep-ph/9903334.
6. M. Frank and H. Hamidian, Phys. Rev. **D54** (1996) 6790; K. Huitu, J. Maalampi, M. Raidal, Nucl. Phys. **B420** (1994) 449.
7. R. N. Mohapatra and A. Rašin, Phys. Rev. **D54** (1996) 5835.
8. M. Frank, H. Hamidian, K. Puolamäki, hep-ph/9903283 (to appear in Phys. Lett. B.); hep-ph/9904458.

TABLE I. $\Lambda_{\text{SUSY}} = 50\text{T eV}$, $\Lambda_M = 10\Lambda_{\text{SUSY}}$, fixed $\tan\beta = 15$.

$(n_3, n_8, n_H,$ $n_L, n_{e^c}, n_{\nu^c})$	μ M_3	$m_{H^\pm}$ $m_{\tilde{\nu}_e}/m_{\tilde{\nu}_\tau}$	$m_{\tilde{\chi}^\pm_{1,2}}$ $m_{\tilde{u}_{1,2}}$	$m_{\tilde{\chi}^0_{1,2}}$ $m_{\tilde{t}_{1,2}}$	$m_{\tilde{e}_{1,2}}$ $m_{\tilde{d}_{1,2}}$	$m_{\tilde{\tau}_{1,2}}$ $m_{\tilde{b}_{1,2}}$
$(1,1,1,0,0,0)$	439	514	346/469	40/346	70/312	54/314
	1060	302/302	1000/1045	905/1019	1001/1048	990/1006
$(1,1,0,1,0,0)$	441	513	347/471	40/347	90/322	78/323
	1060	312/312	999/1045	904/1019	1000/1048	989/1006
$(1,1,1,0,1,0)$	438	514	345/468	122/346	104/318	93/320
	1060	309/308	1001/1045	906/1019	1002/1048	990/1007
$(1,1,0,1,1,0)$	440	514	346/470	122/347	115/327	106/328
	1060	317/317	1000/1045	905/1019	1001/1048	990/1007
$(1,1,1,0,0,1)$	438	514	345/468	40/346	100/318	89/320
	1060	308/308	1001/1045	906/1019	1002/1048	990/1007
$(1,1,0,1,0,1)$	440	513	346/470	40/346	112/326	103/328
	1060	317/316	1000/1045	905/1019	1001/1048	990/1007
$(1,1,1,0,0,0)$	−439	510	354/463	41/353	70/312	54/314
	1060	302/302	1000/1045	908/1016	1001/1048	989/1005
$(1,1,0,1,0,0)$	−441	509	355/465	41/354	90/322	78/323
	1060	312/312	999/1045	907/1016	1000/1048	988/1005
$(1,1,1,0,1,0)$	−438	511	354/462	123/353	104/318	93/320
	1060	309/308	1001/1045	909/1016	1002/1048	989/1006
$(1,1,0,1,1,0)$	−440	510	354/464	123/354	115/327	106/328
	1060	317/317	1000/1045	908/1016	1001/1048	989/1006
$(1,1,1,0,0,1)$	−438	511	354/462	41/353	100/318	90/320
	1060	308/308	1001/1045	909/1016	1002/1048	989/1006
$(1,1,0,1,0,1)$	−440	510	354/464	41/354	112/326	103/328
	1060	317/316	1000/1045	908/1016	1001/1048	989/1006

Ultrastrong Coupling in Supersymmetric Gauge Theories

Alex Buchel

Newman Laboratory, Cornell University, Ithaca NY 14853

Abstract. We study "ultrastrong" coupling points in scale-invariant $N = 2$ gauge theories. These are theories where, naively, the coupling becomes infinite, and is not related by S-duality to a weak coupling point. These theories have been somewhat of a mystery, since in the M-theory description they correspond to points where parallel M 5-branes coincide. Using the low-energy effective field theory arguments we relate these theories to other known $N = 2$ CFT.

INTRODUCTION

Strongly coupled dynamics of gauge theories is a longstanding problem. Not accessible by methods of perturbation theory it requires new approaches but rewards with interesting physics. A typical example is a lattice formulation of $SU(3)$ Yang-Mills theory where color confinement occurs in the strong coupling limit. The present paper addresses the ultrastrong coupling $g \to \infty$ of the scale invariant $N = 2$ supersymmetric field theories. A novel feature compared to the non-supersymmetric models is the existence of a manifold of degenerate vacua which leads to a rich structure of the critical behavior.

Recently, electric-magnetic dualities have played a major role in our understanding of gauge theories and string theories. Such dualities are usually properties of the low-energy effective actions. However, in certain scale invariant theories a particular subgroup of the low-energy dualities is promoted to the exact "symmetry" group — S-duality group — of the full quantum theories. Classical examples are $N = 4$ supersymmetric Yang-Mills theories and finite $N = 2$ theories. Define the microscopic gauge coupling $\tau = \frac{\theta}{\pi} + i\frac{8\pi}{g^2}$. The $N = 4$ gauge theories and $N = 2$ theory with $SU(2)$ gauge group and four fundamental "quark" hypermultiplets have S-duality group isomorphic to $SL(2, \mathbf{Z})$. The classical coupling space of these theories, Im $\tau > 0$, is identified under the transformations $T : \tau \to \tau + 1$ and $S : \tau \to -1/\tau$. Here, S-duality group is big enough to identify all ultrastrong coupling points Im $\tau = 0$ with weak coupling $g = 0$. For the higher rank scale invariant $N = 2$ gauge theories this is not the case. The S-duality group of the scale invariant $SU(n)$ theory, $n > 2$, is isomorphic to $\Gamma^0(2) \subset SL(2, \mathbf{Z})$, generated by $T : \tau \to \tau + 2$ and $S : \tau \to -1/\tau$. Ultrastrong couplings $\tau = \pm 1$, are fixed points of the duality identifications and so they can not be mapped to a weak coupling $\tau = +i\infty$.

The strong coupling behavior of the low energy effective description of $N = 2$ supersymmetric gauge theories has been originally understood for $SU(2)$ QCD in [1] and later generalized to higher rank $SU(n)$ models in [2]. The Coulomb branch of these models is a manifold of generate vacua parameterized at weak coupling by the VEVs of the adjoint scalar in the vector hypermultiplet. At a generic point on the Coulomb branch the $SU(n)$ gauge group is broken to $U(1)^{n-1}$ with effective coupling τ_{ij}. In such vacua there are no massless states charged under the low energy $U(1)$'s and so the effective coupling τ_{ij} does not run in the infrared. Over the special submanifolds of the Coulomb branch the low energy effective coupling becomes singular. This singularity is due to generically massive on the Coulomb branch charged states becoming massless over these special submanifolds. Contrary to the singularities on the Coulomb branch which are at least of complex codimension one, the *whole* Coulomb branch of a scale invariant theory at ultrastrong coupling is singular. This novel codimension zero singularity of the Coulomb branch is yet another motivation to study the ultrastrong coupling.

Finally, the M-theory/IIA realization of $N = 2$ supersymmetric gauge theories on intersecting branes [4] adds even more mystery to the problem of ultrastrong coupling. It turns out that the $g \to \infty$ limit of the scale invariant theory corresponds to the limit of coincident NS 5-branes of Type IIA strings. Whether or not the string construction of $N = 2$ scale invariant models at ultrastrong coupling correctly describes the field theory remains an open question.

SINGULARITIES IN THE MODULI SPACE OF SCALE INVARIANT $SU(3)$ $N = 2$ THEORY

$N = 2$ $SU(3)$ scale invariant theory is described in terms of $N = 1$ superfields by a field strength chiral multiplet W_α and a scalar chiral multiplet Φ both in the adjoint of the gauge group, and 6 chiral multiplets Q^i in the **3** and $\tilde{Q}_i$ in the $\overline{\bf 3}$ representations of the gauge group, where $i = 1, \ldots 6$ is a flavor index. We denote the complex scalar components of Φ, Q and $\tilde{Q}$ and their vevs by the same symbols. Classically, the $N = 2$ superpotential $\mathcal{W}$ is

$$\mathcal{W} = \tau\sqrt{2}\, \tilde{Q}_i^a \Phi_a^b Q_b^i, \qquad (1)$$

where $a, b = 1, \ldots 3$ are color indices and τ is a microscopic coupling. In the Coulomb phase Φ takes values in the Cartan subalgebra of the gauge group with $Q = \tilde{Q} = 0$, implying that Φ can be diagonalized by a color rotation to a complex traceless matrix

$$\Phi = \text{diag}(\phi_1, \phi_2, \phi_3), \qquad \sum_a \phi_a = 0. \qquad (2)$$

This VEV generically breaks the gauge symmetry to $U(1)^2$ and gives masses of order ϕ_a to chiral multiplets (quarks) Q and $\tilde{Q}$. Below the mass scale of the lightest quark, the low-energy effective action is a free $U(1)^2$ $N=2$ gauge theory with effective coupling

$$\tau_{ij} = \frac{1}{2}\tau\, C_{ij}, \qquad (3)$$

where C_{ij} is a Cartan matrix of $SU(3)$. Expression (3) is valid only in a weak coupling limit $\tau \to +i\infty$ and for a generic VEV of Φ. Quantumly, effective coupling τ_{ij} forms a section of an $Sp(4,\mathbb{Z})$ bundle on the Coulomb branch reflecting the EM duality of the low energy effective description. The matrix of the effective coupling was identified in [2] with the period matrix of the genus 2 hyperelliptic curve Σ_3

$$y^2 = \left(x^3 - ux - v\right)^2 - fx^6. \qquad (4)$$

Complex moduli u and v parameterize the Coulomb branch $\mathcal{C}$ and f is a function of the microscopic gauge coupling τ. At weak coupling $\tau \to +i\infty$, $f \sim e^{i\pi\tau}$ and the Coulomb branch moduli u and v are symmetric polynomials of the expectation values of the adjoint scalar

$$\begin{aligned} u &= -(\phi_1\phi_2 + \phi_1\phi_3 + \phi_2\phi_3), \\ v &= \phi_1\phi_2\phi_3. \end{aligned} \qquad (5)$$

The complex structure of Σ_3 degenerates whenever the discriminant of (4) vanishes

$$\mathcal{C}_s: \quad v^6\,(f-1)f^3\,\left(729fv^4 - [27v^2 - 4u^3]^2\right) = 0. \qquad (6)$$

A degeneration is interpreted physically as massive off a singular submanifold $\mathcal{C}_s \subset \mathcal{C}$ charged states becoming massless precisely at $\mathcal{C}_s$. For a generic gauge parameter f, the submanifold $\mathcal{C}_s$ is a union of three branches

$$\mathcal{C}_s = \mathcal{C}_s^{(1)} \cup \mathcal{C}_s^{(2)} \cup \mathcal{C}_s^{(3)}, \qquad (7)$$

with

$$\begin{aligned} \mathcal{C}_s^{(1)}: &\quad v = 0, \\ \mathcal{C}_s^{(2)}: &\quad \frac{4u^3}{27v^2} = 1 + \sqrt{f}, \\ \mathcal{C}_s^{(3)}: &\quad \frac{4u^3}{27v^2} = 1 - \sqrt{f}. \end{aligned} \qquad (8)$$

The simplest singularity is $\{v=0,\ u \neq 0\} \subset \mathcal{C}_s^{(1)}$. At weak coupling we can parameterized this singularity as $\phi_1 = -\phi_2 \neq 0$ and $\phi_3 = 0$. From (1) it becomes evident that the singularity is introduced by integrating out in the effective description 6 massless quarks charges under one of the low-energies $U(1)$. This conclusion is true quantum mechanically as well. Let $v = \mu\, u$ with $\mu \to 0$. Explicitly evaluating the period matrix of (4) we find

$$\tau_{ij} = \begin{pmatrix} \tau_1 & 0 \\ 0 & -\frac{6i}{\pi} \ln \frac{\mu}{u^{1/2}} \end{pmatrix}, \qquad \mu \to 0, \tag{9}$$

where τ_1 is a function of the microscopic gauge parameter f. An effective coupling of the second $U(1)$ implies the existence of 6 light hypermultiplets. The identification of $v = 0$ singularity at finite microscopic coupling with the 6 massless quark singularity at weak coupling should not be surprising since the two can be connected by a holomorphic path in the gauge coupling space without changing the nature of the singularity (the number and charges of the massless states).

Consider the $\mathcal{C}_s^{(2)}$ branch for $\{u \neq 0, v \neq 0\}$. To understand this singularity at weak coupling, $f = 1/\lambda^4$, $\lambda \to \infty$, parameterize the Coulomb branch moduli as follows

$$\begin{aligned} \phi_1 &= \lambda\Lambda + \mu, \\ \phi_2 &= \lambda\Lambda - \mu, \\ \phi_3 &= -2\lambda\Lambda. \end{aligned} \tag{10}$$

In the limit $\lambda \to \infty$ with $\Lambda \gg \mu$ kept finite, the microscopic gauge group is broken at energy scale $\lambda\Lambda$ to $SU(2) \times U(1) \subset SU(3)$ and all quarks get masses of order $\lambda\Lambda$. Upon integrating out massive quarks, we end up below the scale $\lambda\Lambda$ with pure $SU(2)$ Yang-Mill theory with a strong coupling scale of order Λ. The $SU(2)$ is further broken below μ to $U(1) \subset SU(2)$. Really, redefining in (4) $y \to 3\lambda y$ and $x \to x + \lambda\Lambda$ and taking the limit $\lambda \to \infty$, $\Sigma_3 \to \Sigma_2^{\tilde{\Lambda}}$

$$\Sigma_2^{\tilde{\Lambda}}: \qquad y^2 = (x^2 - \mu^2)^2 - \tilde{\Lambda}^4, \tag{11}$$

with $\tilde{\Lambda}^4 = \Lambda^4/9$. In the given parametrization the $\mathcal{C}_s^{(2)}$ branch reads

$$\mathcal{C}_s^{(2)}: \qquad \mu^2 = \tilde{\Lambda}^2, \tag{12}$$

which is a well known monopole singularity of $SU(2)$ Yang-Mills theory. Much like on the $\mathcal{C}_s^{(1)}$ branch, the weak coupling interpretation of the singularity is valid at finite microscopic coupling since there exists a holomorphic path in the gauge parameter space to a weak coupling regime not intersecting $\mathcal{C}_s^{(1)}$ or $\mathcal{C}_s^{(3)}$ branches.

The $\mathcal{C}_s^{(3)}$ branch can be analyzed similar to the $\mathcal{C}_s^{(2)}$ branch. With the same weak coupling parametrization as in the previous case this singularity is interpreted as a dyon singularity of $\Sigma_2^{\tilde{\Lambda}}$

$$\mathcal{C}_s^{(3)}: \qquad \mu^2 = -\tilde{\Lambda}^2. \tag{13}$$

Unless $f = \{0, 1, \infty\}$, any two branches of $\mathcal{C}_s$ intersect at the conformal point $u = v = 0$ where the full $SU(3)$ gauge group remains unbroken. The moduli space singularities we discuss above thus provide full classification. Note that each branch of $\mathcal{C}_s$ is a complex codimension one submanifold of the Coulomb branch.

ULTRASTRONG COUPLING OF SCALE INVARIANT $N = 2$ THEORIES

Singularities in the low-energy effective action of the scale invariant models are not restricted to the moduli space singularities discussed in the previous section. The effective coupling on the Coulomb branch of the scale invariant $SU(n)$ gauge theory with $2n$ flavors of hypermultiplets (quarks) in the fundamental representation of the gauge group is identified with the period matrix of genus $n-1$ Riemann surface Σ_n

$$y^2 = \left(x^n - \sum_{\ell=2}^{n} u_\ell x^{n-\ell}\right)^2 - f x^{2n}, \tag{14}$$

where u_ℓ are Coulomb branch moduli, and f is a function of the microscopic coupling τ. The weak coupling asymptotic $f \sim e^{\pi i \tau}$ as $\tau \to +i\infty$ is the only unambiguous relation between f and τ. The exact functional form $f(\tau)$ depends on the nonperturbative definition of the gauge coupling. We choose following [2]

$$f = -\frac{4\theta_2^4(\tau)\,\theta_4^4(\tau)}{\left(\theta_4^4(\tau) - \theta_2^4(\tau)\right)^2}. \tag{15}$$

Then, $f = 1$ is the ultrastrong coupling point $\tau = \pm 1$, and $f \to \infty$ corresponds to a $\mathbf{Z}_2$ coupling $\tau = i$. The complex structure of Σ_n is regular at $f = \infty$ unless $u_n = 0$. In the latter case the singularity is identical to the moduli space singularity associated with a partial restoration of the original nonabelian gauge symmetry. On the other hand, (14) is singular for any choice of the Coulomb branch moduli when $f = \{0, 1\}$.

$f = 0$ singularity of the low-energy effective action is trivial: at weak coupling, masses of W-bosons of the gauge group from adjoint Higgsing, as well as quark masses are of order $g|\phi_a|$. In the limit $g \to 0$ they become massless causing this singularity.

We now turn to the description of $SU(n)$ scale invariant theories at ultrastrong coupling. Most features are already present in the $SU(3)$ theory which we describe in details. We outline generalization to higher rank $SU(n)$ theories. It is clear that described techniques could be used to study ultrastrong coupling of scale invariant $N = 2$ models with other simple groups and theories with product gauge groups.

While the complex structure of Σ_3 generates over the whole Coulomb branch at $f = 1$, this degeneration depends on the vacuum moduli Fig. 1a-c.

For a generic vacuum $u \neq 0$ and $v \neq 0$ (Fig 1a.) one of the handles of Σ_3 pinches off in the limit $f \to 1$. A physical interpretation of the singularity is facilitated by an appropriate choice of a symplectic homology basis. Choosing the basis with the vanishing cycle representing an electrically charged state under the second $U(1)$, we find

$$\tau_{ij}^{(a)} = \begin{pmatrix} \tau_1 & 0 \\ 0 & -\frac{i}{2\pi}\ln(1-f) \end{pmatrix}, \tag{16}$$

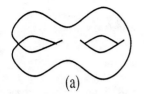 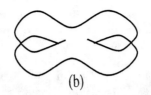 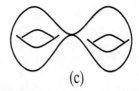

FIGURE 1. Moduli dependent degenerations of the complex structure of the scale invariant $SU(3)$ gauge theory at ultrastrong coupling. Degeneration (a) occurs for $u \neq 0$ and $v \neq 0$. In (b), $u \neq 0$ and $v = 0$. In (c), $u = 0$ and $v \neq 0$.

where τ_1 goes to a constant in the limit $f \to 1$. The degeneration (16) implies that the low-energy $U(1)$ factors decouple. The fact that τ_1 turns to a constant means that the lightest state charged under the first $U(1)$ is massive. The other $U(1)$ becomes free when $f = 1$. The renormalization group running of its coupling is characteristic for an appearance of a massless charged state. A priori it is not clear whether the massless state is in the vector or chiral $N = 2$ multiplet; nor is it possible to fix the number of massless states. We assume for now that there is a single massless state in the chiral multiplet. As we see, this fits nicely into consistent picture with other degenerations of Σ_3.

At the ultrastrong coupling of $u \neq 0$, $v = 0$ vacuum (Fig 1b.) both handles of Σ_3 pinch off. One of the handles degenerates in a same way as for a generic vacuum $u \neq 0$ and $v \neq 0$, and the degeneration of the other one is analogous to the 6 massless quark moduli space singularity discussed in the previous section. Indeed, the period matrix of Σ_3 for $v^2/u^3 \equiv \mu^6/u^3 \to 0$ and $f \to 1$ evaluates to

$$\tau_{ij}^{(b)} = \begin{pmatrix} -\frac{6i}{\pi} \ln \frac{\mu}{u^{1/2}} & 0 \\ 0 & -\frac{i}{2\pi} \ln(1-f) \end{pmatrix}, \qquad (17)$$

implying two decoupled infrared-free $U(1)$'s with the described massless spectrum.

The ultrastrong coupling of $u = 0$ and $v \neq 0$ vacuum (Fig 1c.) is more interesting. In this case the curve (4) "splits" into two tori. Note that the splitting occurs over a zero cycle in the homology. In this case we find

$$\tau_{ij}^{(c)} = \begin{pmatrix} \omega & 0 \\ 0 & -\omega^2 \end{pmatrix}, \qquad (18)$$

where $\omega = e^{2\pi i/3}$. As before, low-energy $U(1)$'s decouple. The non-degenerate form of (18) seems to imply that all charged states should be massive. This conclusion is however incompatible with the existence of massless charged states for generic vacua, as signaled by the degenerate form of $\tau_{ij}^{(a)}$ at $f = 1$. A situation where a massless for $u \neq 0$ state becomes massive at $u = 0$ implies some kind of discontinuous behavior one does not expect in supersymmetric theories. Rather, we assume that the state massless for $u \neq 0$ will remain massless for $u = 0$ as well. Using low-energy EM duality we can choose this state to have electric and magnetic charges correspondingly $(n_e, n_m) = (1, 0)$ with respect to the second $U(1)$. At the given vacuum, an anomaly-free $U(1)_R$ symmetry of the microscopic theory is broken to a discrete $\mathbf{Z}_3$ subgroup by a nonzero value of v. The massless $(1, 0)$ state together with $(0, 1)$ and $(1, 1)$ states form a triplet under the global $\mathbf{Z}_3$. States charged under the first $U(1)$ have mass scale of order $v^{1/3}$. The effective coupling of the other $U(1)$ is independent of the separation of scales $v^{1/3}$ and the scale set by the light states (massless in the limit $f \to 1$). The latter suggests that the second $U(1)$ with mutually nonlocal states $(1, 0)$, $(0, 1)$ and $(1, 1)$ is an interacting conformal field theory. In fact, it is equivalent to the $\mathbf{Z}_3$ Argyres-Douglas point of $N = 2$ $SU(3)$ Yang-Mills theory [3]. The fixed value of the conformal coupling can be understood heuristically as follows. Assume that the renormalization group flow of a theory with mutually nonlocal charges is the sum of flows generated separately by each state. Then, the running of the gauge coupling due to the massless spectrum $(n_e^{(j)}, n_m^{(j)})$ is given by [3]

$$\frac{\partial}{\partial \log \mu} \tau = -\frac{i}{\pi} \sum_j \left[n_m^{(j)} \tau + n_e^{(j)} \right]^2. \tag{19}$$

Equation (19) verifies that $\tau_2 = -w^2$ is a RG flow invariant for the massless spectrum $(1, 0)$, $(0, 1)$ and $(1, 1)$.

In the scale invariant $SU(3)$ model at ultrastrong coupling turning off the u moduli for nonzero v changed the behavior of one of the low-energy $U(1)$'s from a free theory to a nontrivial CFT. In higher rank scale invariant theories at $\tau = +i\infty$ by changing the vacuum moduli one encounters transitions between different interacting CFTs. To illustrate the idea, consider the scale invariant $SU(4)$ model. The Coulomb branch of this theory is described by a genus three curve Σ_4

$$y^2 = (x^4 - u_2 x^2 - u_3 x - u_4)^2 - fx^8, \tag{20}$$

where u_2, u_3, u_4 parameterize a vacuum and the gauge parameter f relates to τ through (15). In $u_2 = u_3 = 0$, $u_4 \neq 0$ vacuum, an anomaly free $U(1)_R$ symmetry is broken to $\mathbf{Z}_4$ by the expectation value of u_4. The period matrix of Σ_4 takes the form

$$\tau_{ij}^{\mathbf{Z}_4} = \begin{pmatrix} i & 0 & 0 \\ 0 & 0 & 0 \\ 0 & 0 & i \end{pmatrix}. \tag{21}$$

The low energy $U(1)$'s again decouple. States charged under the first factor have a mass scale of order $u_4^{1/4}$, the second $U(1)$ is free and the other $U(1)$ is an interacting conformal field theory with the $\mathbf{Z}_4$ global symmetry. The massless spectrum of the model is summarized in Table 1.

From (21), the effective couplings of the free and the conformal $U(1)$ factors are

$$\tilde{\tau}_{ij} = \begin{pmatrix} 0 & 0 \\ 0 & i \end{pmatrix}. \tag{22}$$

TABLE 1. Massless spectrum at a Z_4 vacuum of the scale invariant $SU(4)$ model.

State	Free $U(1)$		Conformal $U(1)$	
	n_1	m_1	n_2	m_2
s	0	1	0	0
h_1	0	0	1	0
h_2	0	0	0	1
h_3	0	1	1	0
h_4	0	1	0	1

As before, this fixed value of the effective coupling can be understood using "rank two" generalization of the renormalization group flow of (19)

$$\frac{\partial}{\partial \log \mu} \tilde{\tau}_{k\ell} = -\frac{i}{\pi} \sum_p \left[m_i^{(p)} \tilde{\tau}_{ik} + n_k^{(p)} \right] \left[m_j^{(p)} \tilde{\tau}_{j\ell} + n_\ell^{(p)} \right]. \tag{23}$$

In (23), p summation goes over the massless spectrum. Consider now turning on the $u_3 = \mu^3$ moduli such that $u_3^4/u_4^3 \ll 1$, while keeping $u_2 = 0$. The complex structure of Σ_4 becomes

$$\tau_{ij}^{u_3} = \begin{pmatrix} \tau_1 & 0 & 0 \\ 0 & -\frac{3i}{\pi} \ln \frac{\mu}{u_4^{1/4}} & 0 \\ 0 & 0 & -w^2 \end{pmatrix}, \tag{24}$$

Effective coupling $\tau_{ij}^{u_3}$ implies three decoupled sectors in the infrared. States charged under the first $U(1)$ have mass of order $u_4^{1/4}$ since $\tau_1 \to i$ as $\mu \to 0$. The second $U(1)$ has three light charged states (with masses of order μ). In the limit $\mu \to 0$ these states are identified with massless states s, h_3 and h_4 of the Z_4 critical behavior. Finally, the third $U(1)$ represents a Z_3 CFT with massless states h_1, h_2 and $h_1 - h_2$. Note that in this case the global symmetry of the nontrivial conformal sector has nothing to do with the original $U(1)_R$ symmetry: the latter is completely broken when both u_3 and u_4 moduli are nonzero.

In all cases considered thus far the infrared conformal sector was either a free $U(1)$ gauge theory or rank one interacting CFT or combination of both. Interacting CFTs of rank two occur first at ultrastrong coupling of the scale invariant $SU(5)$ gauge theory. Here, $U(1) \times U(1)$ CFT with a global $Z_5 \subset U(1)_R$ symmetry has coupling

$$\tau_{ij} = \begin{pmatrix} -\lambda^4 & \lambda + \lambda^3 + \lambda^4 \\ \lambda + \lambda^3 + \lambda^4 & \lambda^2 - \lambda^3 \end{pmatrix}, \tag{25}$$

fixed by the massless spectrum of Table 2. In (25), $\lambda = e^{2\pi i/5}$.

TABLE 2. Massless spectrum of $U(1) \times U(1)$ CFT.

State	Conformal $U(1) \times U(1)$			
	n_1	m_1	n_2	m_2
h_1	1	0	0	0
h_2	0	1	0	0
h_3	0	0	1	0
h_4	0	0	0	1
h_5	1	1	1	1

DISCUSSION

We found a consistent picture of $N = 2$ scale invariant theories at ultrastrong coupling. The low-energy effective description of these theories realizes a large class of RG flow fixed theories with mutually nonlocal charges. By changing the Coulomb branch moduli one induces transitions between different interacting conformal sectors of the effective low-energy description.

ACKNOWLEDGMENTS

The work reported on here was done in collaboration with P. C. Argyres. This work is supported in part by NSF grant PHY-9513717.

REFERENCES

1. N. Seiberg and E. Witten, *Nucl. Phys.* **B431**, 484 (1994).
2. P.C. Argyres, M.R. Plesser, and A.D. Shapere, *Phys. Rev. Lett.* **75**, 1699 (1995).
3. P.C. Argyres and M. R. Douglas, *Nucl. Phys.* **B448**, 93 (1995).
4. E. Witten, *Nucl. Phys.* **B500**, 3 (1997).

Lepton - Chargino Mixing and R-Parity Violating SUSY [1]

Mike Bisset

Department of Physics, Tsinghua University, Beijing, 100084, China

Otto C.W. Kong, Cosmin Macesanu and Lynne H. Orr

Department of Physics and Astronomy, University of Rochester, Rochester, New York 14627

Abstract. We present a study of charged lepton mass matrix diagonalization in R-parity violating SUSY. The case in which the bilinear couplings μ_i have large values is given special attention.

INTRODUCTION

R-parity violating SUSY is a subject which has enjoyed a lot of interest in the past few years. The possibility of mixing between particles and superpartners, allowed in these models, makes for very interesting phenomenology. However, due to the large number of parameters present, this subject is also quite difficult to study. Recently, it has been found (see [1]) that by working in a specific basis (single VEV parametrization) the analysis of the fermionic sector is greatly simplified, without loss of generality. In this basis, the only non-MSSM parameters that play a role in the leptonic phenomenology at tree level are the three RPV bilinear couplings μ_i.

The phenomenological consequences of this model in the fermionic sector have been extensively studied [2]. This paper aims to detail the technical aspects concerning the chargino-lepton mass matrix diagonalization. It has been found that for large values of the couplings μ_i (of order of hundreds GeV and above) this problem is not trivial. We should point out that our interest in this issue *here* is mostly theoretical: most of the range of μ_i values relevant to the discussion below is not allowed by the experimental constraints (except in special cases; for more details on this subject see [2]).

The framework is the same as in [1]. The mass matrix can be written:

---

[1] Presented by C. Macesanu

$$M_c = \begin{bmatrix} M_2 & g_v & 0 & 0 & 0 \\ g'_v & \mu_0 & 0 & 0 & 0 \\ 0 & \mu_1 & \bar{m}_1 & 0 & 0 \\ 0 & \mu_2 & 0 & \bar{m}_2 & 0 \\ 0 & \mu_3 & 0 & 0 & \bar{m}_3 \end{bmatrix}$$

where $g_v = g_2 v_u/\sqrt{2} = \sqrt{2} M_w \sin\beta$, $g'_v = g_2 v_d/\sqrt{2} = \sqrt{2} M_w \cos\beta$. Here, by $\bar{m}_i$ we denote the Yukawa masses of the three leptons e, μ, τ. The physical masses will be denoted by m_i.

The aim is to go from the weak interaction fields, in terms of which the lagrangian is written initially, to the mass eigenstate (physical) fields, which can be observed experimentally. To this purpose, we rotate the left fields by a matrix U^L and the right fields by a matrix U^R. These rotations will also diagonalize the squared mass matrices:

$$U^{L\dagger} M^L U^L = U^{R\dagger} M^R U^R = \mathrm{diag}\{M^2_{\chi_1}, M^2_{\chi_2}, m_i^2\}$$

where $M^L = M_c^\dagger M_c$, $M^R = M_c M_c^\dagger$.

The mixing between leptons and charginos naturally leads to changes in the couplings of these particles to the gauge bosons (see [3]). In the case of Z coupling we will have :

$$A^L_{ij} = \tilde{A}^L_{ij} + (1 - 2\sin^2\theta_W)\delta_{ij}, \quad \tilde{A}^L_{ij} = U^L_{i1} U^L_{j1}$$

$$A^R_{ij} = \tilde{A}^R_{ij} - 2\sin^2\theta_W \delta_{ij}, \quad \tilde{A}^R_{ij} = 2U^R_{i1} U^R_{j1} + U^R_{i2} U^R_{j2}$$

The diagonal (δ_{ij}) terms are the SM values, while the $\tilde{A}$ terms are consequences of the mixing. Being nondiagonal, they lead to anomalous Z couplings and non-standard decays (e.g. $Z \to e\mu$, $\mu \to eee$). It turns out that, in most cases, the anomalous left coupling is the important one; the anomalous right coupling is proportional with the product $m_i m_j$ and very small numerically. In what follows we will concentrate on the left rotation matrix.

I ANALYTIC DIAGONALIZATION OF MASS MATRIX

In this section we will analyze approximate analytical solutions to our diagonalization problem. The perturbative solution for small μ's has been given in [1]; we present it here for comparison with further results:

$$U^L_{i1} = \mu_i \frac{\sqrt{2} M_W \cos\beta}{\Delta} \quad , \quad i = e, \mu, \tau \tag{1}$$

$\Delta = \mu_0 M_2 - g_v g'_v$. This formula reveals to us one important fact: the strength of the mixing decreases at large $\tan\beta$. The phenomenological consequences of this behaviour have been analyzed in [1], [2].

The fact that the above formula doesn't work at large μ is apparent; indeed, as we increase μ_i, the components U^L_{i1}, as given by the above formulae, increase indefinitely, which is not allowed by U matrix unitarity.

It is possible to diagonalize the matrix without assuming that the μ_i are small. Instead, we will take the Yukawa masses to be small; actually, we will take them to be zero in the matrix M^L. Then, we get the following solution:

$$U^L = \begin{bmatrix} x_1 & x'_1 & \mu_1 \frac{g'_v}{\Delta_e} & \mu_2 \frac{g'_v \Delta}{\Delta_\mu \Delta_e} & \mu_2 \frac{g'_v \Delta}{\Delta_\tau \Delta_\mu} \\ x_2 & x'_2 & -\mu_1 \frac{M_2}{\Delta_e} & -\mu_2 \frac{M_2 \Delta}{\Delta_\mu \Delta_e} & -\mu_2 \frac{M_2 \Delta}{\Delta_\tau \Delta_\mu} \\ x\frac{\mu_1}{\mu_5} & x'\frac{\mu_1}{\mu_5} & \frac{\Delta}{\Delta_e} & -\frac{\mu_2}{\Delta_\mu}\frac{\mu_1 \alpha_2}{\Delta_e} & -\frac{\mu_3}{\Delta_\tau}\frac{\mu_1 \alpha_2}{\Delta_\mu} \\ x\frac{\mu_2}{\mu_5} & x'\frac{\mu_2}{\mu_5} & 0 & \frac{\Delta_e}{\Delta_\mu} & -\frac{\mu_3}{\Delta_\tau}\frac{\mu_2 \alpha_2}{\Delta_\mu} \\ x\frac{\mu_3}{\mu_5} & x'\frac{\mu_3}{\mu_5} & 0 & 0 & \frac{\Delta_\mu}{\Delta_\tau} \end{bmatrix} \quad (2)$$

where $\Delta^2_e = \Delta^2 + \alpha_2 \mu^2_1$, $\Delta^2_\mu = \Delta^2 + \alpha_2(\mu^2_1 + \mu^2_2)$, $\Delta^2_\tau = \Delta^2 + \alpha_2(\mu^2_1 + \mu^2_2 + \mu^2_3)$, and $\alpha_2 = M^2_2 + g'^2_v$. The first two columns correspond to the two charginos, while the last three columns correspond to the leptons (e, μ, τ respectively).

Note that, in the approximation used, the physical masses of the particles (eigenvalues of the matrix M^L) are also zero; as a consequence, the three lepton eigenvectors are degenerate. To get the correct combination, we employ a limiting procedure: start from the exact eigenvectors, and let the electron mass, muon mass, and tau mass go to zero, in this order. If the lepton mass hierarchy holds also for the Yukawa masses ($\bar{m}_e \ll \bar{m}_\mu \ll \bar{m}_\tau$) it can be shown that this way we get the correct result (2).

Besides lepton eigenvectors, other quantities of interest obtained through diagonalization of the mass matrix are chargino masses. In the small Yukawa mass approximation, we get

$$M_{\chi 1,2} = \frac{\alpha_1 + \alpha_2}{2} \pm \sqrt{(\alpha_1 - \alpha_2)^2 + 4(M_2 g_v + \mu_0 g'_v)^2}$$

with $\alpha_2 = \mu^2_0 + g^2_v + \mu^2_5$ ($\mu^2_5 = \mu^2_1 + \mu^2_2 + \mu^2_3$). Interpretation of the quantities α_1 and α_2 is straightforward; at large μ_5, the mass of the heavier chargino is $M_{\chi 1} \approx \sqrt{\alpha_1} \approx \mu_5$, while the mass of the lighter chargino is $M_{\chi 2} \approx \sqrt{\alpha_2}$. Actually, it can be shown that the lighter chargino mass increases monotonically from the MSSM value (for $\mu_5 = 0$) to $\sqrt{\alpha_2}$ (for $\mu_5 \to \infty$). This behaviour has important phenomenological consequences. Consider the fact that the lower limit on the lighter chargino mass $M_{\chi 2} > 90$ GeV eliminates part of the (M_2, μ_0) plane in the MSSM. With R-parity violating terms, you can expect that this excluded region will shrink; indeed, if we make μ_5 big enough, it might potentially go away completely. The fact that $M_{\chi 2}$ is limited above by $\sqrt{\alpha_2}$ means that some region in the (M_2, μ_0) plane does in fact remain excluded, no matter how strong the R-parity is violated. This region is given by the equation:

$$\sqrt{\alpha_2} < 90 \text{ GeV}$$

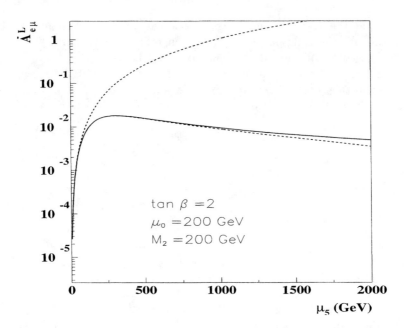

FIGURE 1. Strength of left-handed $Ze\mu$ anomalous coupling : dotted line: small μ approximation (1); solid line: small Yukawa mass approximation (2); dashed line: exact numerical result; μ_i ratio 1:1:1

or, at large $\tan\beta$, $M_2 < 90$ GeV. This result is supported by the exact numerical analysis presented in [1], [2].

Let's turn back to the lepton eigenvectors and consider the anomalous lepton-lepton-Z couplings. For simplicity, let's take the left-handed $Ze\mu$ coupling:

$$\tilde{A}^L_{e\mu} = \mu_1 \mu_2 \frac{g'^2_v}{\Delta^2_e} \frac{\Delta}{\Delta_\mu} \tag{3}$$

At small μ's, the strength of the coupling increases with $\mu_1\mu_2$, while at large μ's it decreases like $1/\sqrt{\mu_5^2}$ (see Fig. 1). In between it will reach a maximum value:

$$\tilde{A}^L_{e\mu\ max} = C\ \frac{g'^2_v}{M_2^2 + g'^2_v}$$

C being a constant which depends only on the ratio μ_1/μ_2. Experimental constraints on anomalous branching ratios or lepton number violating decays (which can generally be written in the form $\tilde{A}^L_{e\mu} < A_0$) will then be satisfied not only in the region of small μ_5, but also for large values of μ_5. Moreover, if in some region of parameter space $\tilde{A}^L_{e\mu\ max} < A_0$ for some process, then that particular process won't contribute at all to constraints on μ_5 values.

The analytic solution (2) derived in this section is not valid at arbitrarily large μ's. The approach used to derive the lepton eigenvectors works only as long as $\bar{m}_e << \bar{m}_\mu << \bar{m}_\tau$. Once we get close to the boundaries of the region where the diagonalization problem has solutions (see next section) this relation does not hold anymore. However, the numerical results for the anomalous couplings $\tilde{A}^L_{ij}$ show an even steeper decrease in this region than that given by (3). As a consequence, for sufficiently large μ_5, constraints from electroweak processes like lepton or Z decay dissapear. This region of large μ_5 is excluded on the basis of strong interaction processes (π decay, or neutrinoless beta decay). However, the corresponding case involving mainly large μ_3 may be phenomenologically viable [2].

II   YUKAWA MASSES

To be able to perform an exact diagonalization of the mass matrix, we first have to find the Yukawa masses of the leptons. Note that this is not the standard problem, in which we have a matrix and we have to find eigenvalues. In this case, we know three of the eigenvalues (the physical masses m_i of the leptons) and we have to find some elements of the matrix itself (the Yukawa masses $\bar{m}_i$).

The direct approach would be to use the eigenvalue equations:

$$\det(M - \lambda I) = 0 \ , \quad \lambda = m_e^2, m_\mu^2, m_\tau^2 \tag{4}$$

This is a three by three system of nonlinear equations, and is not easy to solve even numerically (except in particular cases). So, we will use another approach.

This approach is based on the observation that if we know the solution for some values μ_i^2, we can find the solution at an neighbouring point $\mu_i^2 + \mathrm{d}\mu_i^2$. For simplicity, let's consider the ratio of the μ's fixed, and their magnitude given by a parameter t:

$$\mu_i = r_i \sqrt{t}$$

Write the eigenvalue equations:

$$E(t, \bar{m}_i^2, \lambda) = \det(M - \lambda I) = 0$$

To an infinitesimal modification in t will correspond an infinitesimal modificaton in the Yukawa parameters $\bar{m}_i$:

$$\frac{\partial E}{\partial t} + \frac{\partial E}{\partial \bar{m}_i^2} \frac{\mathrm{d}\bar{m}_i^2}{\mathrm{d}t} = 0 \tag{5}$$

Now, we can numerically integrate this system of linear equations from zero to whatever value of t we need.

What about the existence of solutions for our problem? Let's suppose we can solve the system (4); in order that the result make sense, we require the solutions to be real. This will restrict the allowable values of μ_i to some domain D in μ_i space. What this means, in terms of solutions derived with the help of (5), is that we can increase t only as long as we stay inside this domain. When we reach its boundary, usually what happens is that the determinant of the system (5) becomes zero, and we cannot solve for $d\bar{m}_i$.

Another relevant question is if this domain is simply-connected; that is, starting from $\mu_i = 0$, can we reach any point of it with a path formed by connected straight lines? In other words, does integrating the system (5) allow us access to all the solutions to (4)?

We do not know the answer to this question for the general case of three leptons. But, if we consider the simpler case of only two leptons (presume one of the μ_i is zero), we can write (4) in the form:

$$\begin{cases} a_1 \bar{m}_1^2 + a_2 \bar{m}_2^2 = s \\ a_1 a_2 \bar{m}_1^2 \bar{m}_2^2 = p \end{cases}$$

with solution:

$$(a\bar{m}^2)_{1,2} = \frac{s}{2} \mp \sqrt{s^2 - 4p}$$

The quantity $A = s^2 - 4p$ becomes 0 for some value $\mu_5 = \mu_{5\ max}$ (which gives us the boundary for the domain D), and it can be shown that only for $\mu_5 < \mu_{5\ max}$ is A positive (necessary condition for real solutions). Note that $\mu_{5\ max}$ is generally around a few TeV.

Another interesting issue is the problem of lepton mass hierarchy in this model. In the Standard Model (or MSSM) we have $\bar{m}_e \ll \bar{m}_\mu \ll \bar{m}_\tau$. These relations need not hold in our R-parity violating scenario. Take, for example, the two lepton mixing case presented above. If $\mu_1 = \mu_2$, then, at $\mu_5 = \mu_{5\ max}$, we have $\bar{m}_1 = \bar{m}_2$ (this will happen at quite large μ_5 values, though; $\mu_{5\ max} \sim m_2/m_1$). If $\mu_1 > \mu_2$, it is even possible to get $\bar{m}_1$ greater than $\bar{m}_2$. The next question is if this behaviour holds for the general case of three lepton mixing. The possibility of finding points in parameter space where the three Yukawa masses are of the same order of magnitude (or maybe even equal) is particularly interesting. Unfortunately, the technical difficulties encountered in working with the nonlinear system (4) have stopped us from getting an answer to this question so far.

III CONCLUSIONS

For large R-parity violating terms, the mixing between charginos and charged leptons has different characteristics than at small μ_i. We have presented approximate analytical expressions for both regimes, which can help to understand numerical results.

The problem of existence of solutions of the system (4) - finding the Yukawa masses so that three of the mass matrix eigenvalues will be equal to the physical lepton masses - is still unsolved for the general case of three lepton mixing (although within the phenomenologically viable region, numerical solutions are always successfully obtained in [2]). For two lepton mixing it can be solved, and it has been shown that there are regions in parameter space where the Yukawa masses of the two particles are of the same order of magnitude.

REFERENCES

1. M. Bisset, O.C.W. Kong, C. Macesanu, L.H. Orr, Phys. Lett. **B430**:274-280, 1998 (hep-ph/9804282)
2. M. Bisset, O.C.W. Kong, C. Macesanu, L.H. Orr, (hep-ph/9811498)
3. M. Novakowsky, A. Pilaftsis, Nucl. Phys. **B461**:19-49, 1996

Flavour Changing Neutral Currents in Supersymmetric Models with Large $\tan\beta$[1]

C. Hamzaoui*, M. Pospelov†, A. Raymond* and M. Toharia*

*Département de Physique, Université du Québec à Montréal
C.P. 8888, Succ. Centre-Ville, Montréal, Québec, Canada, H3C 3P8

†Theoretical Physics Institute, 431 Tate Laboratory of Physics
Minneapolis, Minnesota 55455

Abstract. We show that in supersymmetric models with nontrivial flavour structure in the soft-breaking sector, one can have flavour changing neutral currents mediated by neutral Higgses. For $\Delta S, \Delta B = 2$ processes and for a generic form of the soft-breaking terms, we find that Higgs-mediated FCNC amplitudes increase very rapidly with $\tan\beta$ and can exceed $SUSY$ box contribution by up to two orders of magnitude when $\tan\beta \sim m_t/m_b$.

I INTRODUCTION

Flavour Changing Neutral Currents (FCNC) processes constitute a good arena for testing and constraining models of fundamental interactions. These flavour-changing processes contribute to many CP-conserving and CP-violating physical quantities. Of most interest of these quantities are the mixing parameters and rare decay rates for the K and B mesons. In this talk, we focus specifically on the $\Delta S, \Delta B = 2$ processes mediated by Higgs particles in supersymmetric models paying special attention to the large $\tan\beta$ regime [1].

The contribution from the soft-breaking sector to the mass splitting in neutral K and B mesons is commonly attributed to the box diagram with the superpartners inside and nontrivial flavour structure of the scalar quark mass matrices [2,3]. We show that there is another class of diagrams which is related to the supersymmetric corrections to the Higgs potential. These corrections change the familiar Yukawa interaction generated by the superpotential making it similar to generic two-Higgs doublet type of interaction with the presence of H_u^* and H_d^* fields. It is clear that this two-doublet model will possess FCNC mediated by neutral Higgses *if*

---

[1] Talk given by M.T.

the squark sector has nontrivial flavour structure. The purpose of this talk is to establish how large these Higgs-mediated FCNC amplitudes can be.

II HIGGS-MEDIATED FCNC AND THE LIMITS ON THE SOFT-BREAKING SECTOR

Below the supersymmetric threshold the Yukawa interaction of the quarks with the Higgs fields has the generic form of the two-Higgs doublet model [4]:

$$-\mathcal{L}_Y = \overline{U}_R Q_L H_u (\mathbf{Y}_u^{(0)} + \mathbf{Y}_u^{(1)}) + \overline{U}_R Q_L H_d^* \mathcal{Y}_u + \overline{D}_R Q_L H_d (\mathbf{Y}_d^{(0)} + \mathbf{Y}_d^{(1)}) \quad (1)$$
$$+ \overline{D}_R Q_L H_u^* \mathcal{Y}_d + h.c.,$$

and $SU(2)$ indices are suppressed here.

The radiative corrections are of the order of the tree level contribution if $\tan \beta$ is large and $\mu \sim m_{sq} \sim m_\lambda$. The diagrams generating $\mathcal{Y}_d$ are shown in Fig. 1 and they also generate the corrections to the mixing angles if flavour can be changed on the squark line.

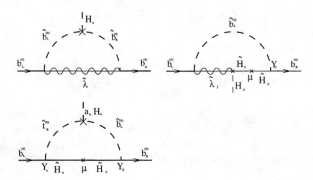

FIGURE 1. The diagrams generating v_u-dependent threshold corrections to $\mathbf{M_d}$.

The presence or absence of Higgs-mediated FCNC in the Down quark sector depends on the particular forms of the matrices $\mathbf{Y}_d^{(0)}, \mathbf{Y}_d^{(1)}$ and $\mathcal{Y}_d$. But even before analyzing possible forms of the mass matrices and Yukawa couplings, we are able to present four-fermion operators which can produce leading effects [5].

The bi-unitary transformation which diagonalizes the Down quarks mass matrix brings also the off-diagonal entries into the four-fermion interaction:

$$\begin{aligned}-\mathcal{L}_{4f} = &\, \overline{D}_{Li} D_{Rj} \overline{D}_{Rk} D_{Ll} |y_b^{(0)}|^2 V_{3i}^* U_{3j} U_{3k}^* V_{3l} \frac{2}{m_A^2} \\ &+ \frac{1}{2} \overline{D}_{Li} D_{Rj} \overline{D}_{Lk} D_{Rl} |y_b^{(0)}|^2 V_{3i}^* U_{3j} V_{3k}^* U_{3l} \frac{1}{m_h^2 \tan^2 \beta} \\ &+ \frac{1}{2} \overline{D}_{Ri} D_{Lj} \overline{D}_{Rk} D_{Ll} |y_b^{(0)}|^2 U_{3i}^* V_{3j} U_{3k}^* V_{3l} \frac{1}{m_h^2 \tan^2 \beta}\end{aligned} \quad (2)$$

Performing standard QCD running of $\bar{q}_L q_R \bar{q}_R q_L$ operator from 500 GeV down to the scale of 5 GeV and 1 GeV and taking matrix elements in vacuum insertion approximation, we arrive at the following formulae for mass splitting in the neutral B-mesons, K-mesons and ϵ_K parameter (see Table 1)

Δm_B	$\left(\frac{500 \text{ GeV}}{m_H}\right)^2 |y_b^{(0)}|^2 |V_{31}^* U_{33} V_{33} U_{31}^*|$	$< 1.1 \cdot 10^{-7}$
Δm_K	$\left(\frac{500 \text{ GeV}}{m_H}\right)^2 |y_b^{(0)}|^2 |V_{31}^* U_{32} V_{32} U_{31}^*|$	$< 1.2 \cdot 10^{-9}$
ϵ_K	$\left(\frac{500 \text{ GeV}}{m_H}\right)^2 |y_b^{(0)}|^2 \text{Im}\,(V_{31}^* U_{32} V_{32} U_{31}^*)$	$< 3.5 \cdot 10^{-12}$

Table 1

If $y_b^{(0)} \sim y_t^{(0)} \simeq 1$ (large $\tan\beta$ regime), it implies very strong constraints on the off-diagonal elements of V and U. These constraints are 'oblique', i.e. the specifics of the soft-breaking sector in different supersymmetric models enters only through V_{13}, U_{23}^*, etc. They are strongly violated if we take $|y_b^{(0)}| \simeq 1$, $|V_{31}| \sim |U_{31}| \sim |V_{ts}|$ and $|V_{32}| \sim |U_{32}| \sim |V_{td}|$. It shows that unlike the eigenvalues of the mass matrices, the mixing angles should not receive more than 10% renormalization from the threshold corrections if both V and U matrices are nontrivial. The numbers on the r.h.s. in Table 1 coincide with the numbers quoted usually as the limits on the "superweak interaction" [6]. Moreover, the tightest constraint from ϵ_K suggests the possibility of having nearly real V_{KM} with CP-violation in the Kaon sector coming from the small phases of the order 10^{-2} in V and U.

For the generic form of the soft-breaking sector, the limits from Table 1 can be converted into the limits on the off-diagonal entries in the Down squark mass matrices $\mathbf{M}_Q^2$ and $\mathbf{M}_D^2$. Treating these entries as the mass insertions on the scalar quark lines and taking also $(\mathbf{M}_Q^2)_{ii} \sim (\mathbf{M}_D^2)_{ii} \sim m_{\tilde\lambda}^2 = m^2$, we are able to calculate V_{ij} and U_{ij} [5]

$$V_{3i} = \frac{1}{3}\frac{1}{3\pi}\frac{\alpha_s v}{(v_d + (\mu/m)(\alpha_s/3\pi)v)}\frac{\mu}{m}\frac{(\mathbf{M}_Q^2)_{3i}}{m^2} \qquad (3)$$

$$U_{3i} = \frac{1}{3}\frac{1}{3\pi}\frac{\alpha_s v}{(v_d + (\mu/m)(\alpha_s/3\pi)v)}\frac{\mu}{m}\frac{(\mathbf{M}_D^2)_{3i}}{m^2}$$

Here we keep only gluino exchange contribution and neglect diagrams with charginos and neutralinos. We summarize the constraints on the $SUSY$ parameter space for the case of the generic form of the soft-breaking in Table 2

Δm_B	$\left(\frac{500 \text{ GeV}}{m_H}\right)^2 |y_b^{(0)}|^4 \left|\frac{\mu}{m}\right|^2 \frac{(\mathbf{M}_{d_L}^2)_{31}}{m^2}\frac{(\mathbf{M}_{d_R}^2)_{31}}{m^2}$	$< 1.4 \cdot 10^{-6}$
Δm_K	$\left(\frac{500 \text{ GeV}}{m_H}\right)^2 |y_b^{(0)}|^6 \left|\frac{\mu}{m}\right|^4 \frac{(\mathbf{M}_Q^2)_{31}}{m^2}\frac{(\mathbf{M}_D^2)_{31}}{m^2}\frac{(\mathbf{M}_Q^2)_{32}}{m^2}\frac{(\mathbf{M}_D^2)_{32}}{m^2}$	$< 2.0 \cdot 10^{-7}$
ϵ_K	$\left(\frac{500 \text{ GeV}}{m_H}\right)^2 |y_b^{(0)}|^6 \text{Im}\left(\left(\frac{\mu}{m}\right)^4 \frac{(\mathbf{M}_Q^2)_{31}}{m^2}\frac{(\mathbf{M}_D^2)_{31}}{m^2}\frac{(\mathbf{M}_Q^2)_{32}}{m^2}\frac{(\mathbf{M}_D^2)_{32}}{m^2}\right)$	$< 5.5 \cdot 10^{-10}$

Table 2

We take m_λ as the real parameter and therefore we have to include possible phase of μ in the ϵ_K-constraint. The fourth and sixth power of $y_b^{(0)}$ in Table 2 at low and intermediate $\tan\beta$ correspond to the $\tan^4\beta$ and $\tan^6\beta$ growth of Higgs-mediated FCNC amplitudes [5].

The constraints on the soft-breaking terms quoted in Table 2 are very sensitive to the value of $y_b^{(0)}$ and μ/m ratio. Nevertheless, they provide valuable limits on the squark flavour sector in the case of large $\tan\beta$, complementary and sometimes much stronger than those from the box diagram if $y_b^{(0)} \sim \mu/m \sim 1$ [3]. The exception are the limits on $(\mathbf{M}_Q^2)_{12}$, $(\mathbf{M}_D^2)_{12}$ entries in the squark mass matrices. The limits on these entries provided by Higgs exchange are relaxed by some power of the ratio $y_s^{(0)}/y_b^{(0)}$ and we do not quote them here.

Many specific *SUSY* models predict certain patterns for the squark mass matrices so that FCNC amplitudes can be calculated in more details and where the comparison with the box diagram contribution can be made.

III CONSTRAINTS ON THE SOFT-BREAKING TERMS IN DIFFERENT SUSY MODELS

In what follows we consider the cases of Minimal supergravity model, *SUSY LR* and *SUSY SO*(10) models and so called "effective supersymmetry".

1. Minimal supergravity model

It is customary to assume, at the scale of the breaking, that the following, very restrictive conditions are fulfilled:

$$\mathbf{M}_Q^2 = m_Q^2 \mathbf{1}; \quad \mathbf{M}_D^2 = m_D^2 \mathbf{1}; \quad \mathbf{M}_U^2 = m_U^2 \mathbf{1} \quad \text{"degeneracy"} \tag{4}$$

$$\mathbf{A}_u = A_u \mathbf{Y}_u; \quad \mathbf{A}_d = A_d \mathbf{Y}_d \quad \text{"proportionality"}. \tag{5}$$

These conditions, if held, ensure that the physics of flavour comes entirely from the superpotential. We would refer to this possibility as to the supergravity scenario. Further RG evolution of the soft-breaking parameters, from the scale of the *SUSY* breaking down to the electroweak scale, induces significant off-diagonal terms in $\mathbf{M}_Q^2$ whereas $\mathbf{M}_D^2$ and $\mathbf{M}_U^2$ stay essentially flavour blind. As a result, no significant right-handed rotation angles can be generated at the threshold, i.e. $|U_{ij}| \sim \delta_{ij}$. The left-handed squark mass matrix is nontrivial, though, and, as a result, 13 and 23 elements of the rotation matrix V are proportional to V_{td} and V_{ts} times the characteristic splitting in left-handed sector induced by RG running, $y_t^2 3\ln(\Lambda/m)/(8\pi^2)$. Thus, the Higgs exchange would induce the operator $\bar{d}_L b_R \bar{d}_L b_R$ proportional to V_{td}^2 which is relevant for the B-meson splitting. Unfortunately, in the case of the Higgs exchange this operator is suppressed by $\tan^{-2}\beta$ (See Eq. (2)) and cannot compete with the box diagram. For the $\Delta S = 2$ processes the degree of suppression is even higher, since the Higgs mediation mechanism involves $\left(y_s^{(0)}/y_b^{(0)}\right)^2$.

2. Effective supersymmetry

The departure from the strict conditions of degeneracy and proportionality may occur in several ways which usually implies significant $SUSY$ contributions to FCNC amplitudes. Here we would skip the theoretical justification for certain choices of the soft-breaking parameters going directly to the phenomenological consequences related with Higgs-mediated FCNC.

We turn now to the "effective supersymmetry" picture [7] which has much less degrees of freedom and where flavour physics can be formulated in a more definitive way. In the squark mass matrices, diagonalized by unitary transformations $\tilde{U}$ and $\tilde{V}$, the eigenvalues corresponding to the first and second generation of squarks, m_1^2, m_2^2, $m_1'^2$, $m_2'^2$, are taken to be in the multi-TeV scale and eventually decoupled from the rest of particles. At the same time, the squarks from the third generation are believed to be not heavier than 1 TeV and weakly coupled to the first and second generations of quarks to avoid the excessive fine-tuning in the radiative corrections to the Higgs potential and suppress FCNC contribution to the Kaon mixing. The advantage of this approach in our case is that the same loop integral stands for the renormalization of the Yukawa couplings and mixing angles [8]. Moreover, it is easy to see that matrices $\tilde{V}$, $\tilde{U}$, V and U are closely related [5].

Table 3 summarizes the limits on the parameter space of the model, provided by $\Delta F = 2$ amplitudes mediated by Higgses

Δm_B	$\left(\frac{500 \text{ GeV}}{m_H}\right)^2 \lvert y_b^{(0)}\rvert^4 \left\lvert\frac{\mu}{m_3}F(m_\lambda/m_3)\right\rvert^2 \left\lvert\tilde{V}_{31}\tilde{U}_{31}\right\rvert < 1.6 \cdot 10^{-7}$
Δm_K	$\left(\frac{500 \text{ GeV}}{m_H}\right)^2 \lvert y_b^{(0)}\rvert^6 \left\lvert\frac{\mu}{m_3}F(m_\lambda/m_3)\right\rvert^4 \left\lvert\tilde{V}_{31}^*\tilde{U}_{32}\tilde{V}_{32}\tilde{U}_{31}^*\right\rvert < 2.4 \cdot 10^{-9}$
ϵ_K	$\left(\frac{500 \text{ GeV}}{m_H}\right)^2 \lvert y_b^{(0)}\rvert^6 \text{Im}\left(\left(\frac{\mu}{m_3}F(m_\lambda/m_3)\right)^4 \tilde{V}_{31}^*\tilde{U}_{32}\tilde{V}_{32}\tilde{U}_{31}^*\right) < 7.0 \cdot 10^{-12}$

Table 3

$$\text{where} \quad F(x) = \frac{2x}{1-x^2} + \frac{2x^3 \ln x^2}{(1-x^2)^2}, \quad \text{with} \ F(1) = 1 \tag{6}$$

In Fig. 2 we plot F and F^4 as a function of the ratio m_λ/m_3. It is interesting to note that Higgs-mediated amplitudes can probe the region of the parameter space where gluino is considerably heavier than stop and sbottom. All other observables related with dipole amplitudes drop off very fast with m_λ and do not produce significant constraints in this part of the parameter space. The comparison of the box diagram contribution to the $B - \bar{B}$ mixing with the Higgs-mediated mechanism provides the value of the critical $(\tan\beta)_{cr}$ above which the Higgs exchange dominates:

$$(\tan\beta)_{cr} = 17 \tag{7}$$

The numbers from Table 3 suggest also that the observable value of ϵ_K parameter in this model can be achieved through the small imaginary phases of μ and/or $\tilde{V}_{ij}$, $\tilde{U}_{ij}$.

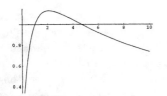 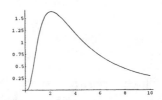

FIGURE 2. In the case of decoupled first and second generation of squarks, the invariant functions F and F^4 are plotted against m_λ/m_3 ratio. The Higgs-mediated FCNC amplitudes are sensitive to large m_λ.

3. Supersymmetric SO(10) and left-right models

Another example from $SUSY$ theory that we would like to consider here are the supersymmetric $SO(10)$ [9,10] and the left-right models [11]. In $SO(10)$ model, left- and right-handed quark superfields are unified at some scale Λ_{GUT} into the same multiplet. Therefore, above the scale of the unification the RG evolution of $\mathbf{M}_Q^2$ and $\mathbf{M}_D^2$ should be the same and these matrices will be equally affected by large Yukawa couplings of the third generation. Therefore, if $\mathbf{M}_Q^2$ exibits non-trivial flavour structure, so does $\mathbf{M}_D^2$. As a result, the squark degeneracy condition is violated both in the left- and right-handed sector and the characteristic splitting is proportional to $y_t^2 \ln(\Lambda_{Plank}^2/\Lambda_{GUT}^2)$. In the basis where $\mathbf{Y}_d^{(0)}$ matrix is diagonal, the off-diagonal entries in the squark mass matrices can be related to the mixing angles of the rotation matrices between two basis given by diagonal $\mathbf{Y}_d^{(0)}$ and diagonal $\mathbf{Y}_u^{(0)}$. Further RG evolution, from Λ_{GUT} to the weak scale, deviates $\mathbf{M}_Q^2$ from $\mathbf{M}_D^2$ but the nontrivial part of $\mathbf{M}_D^2$ survives and generates very important phenomenological consequences. It is easy to see that we can have the same phenomenology in left-right models [13]. The role of Λ_{GUT} in this case is played by Λ_{LR}, the scale where left-right symmetry gets spontaneously broken and heavy right-handed gauge bosons decouple.

To study the Higgs-mediated FCNC amplitudes, we need to know the initial form of the Yukawa matrix $\mathbf{Y}_d^{(0)}$ in the basis where $\mathbf{Y}_u^{(0)}$ is diagonal. Two choices seem to be justified: hermitian $\mathbf{Y}_d^{(0)}$ [12] or complex symmetric $\mathbf{Y}_d^{(0)}$ [14]. Depending on this choice, right-handed mixing matrix is either exactly the same as KM matrix or equal to the transposed KM matrix times the diagonal matrix with two new physical phases in it. The relations between the mixing angles are given in [5].

Injecting these relations into the general limits from Table 1, we obtain the following constraints on the parameter space of the $SO(10)$ and left-right symmetric types of models:

Δm_B	$\left(\frac{500 \text{ GeV}}{m_H}\right)^2 \lvert y_b^{(0)} \rvert^4 \left(\frac{\mu}{m_3}G\right)^2$	$< 1.6 \cdot 10^{-3}$
Δm_K	$\left(\frac{500 \text{ GeV}}{m_H}\right)^2 \lvert y_b^{(0)} \rvert^6 \left(\frac{\mu}{m_3}G\right)^4$	$< 1.5 \cdot 10^{-2}$
ϵ_K	$\left(\frac{500 \text{ GeV}}{m_H}\right)^2 \lvert y_b^{(0)} \rvert^6 \left(\frac{\mu}{m_3}G\right)^4 \sin(\phi_s - \phi_d)$	$< 4.4 \cdot 10^{-4}$

Table 4

where $G(m_\lambda, m_3, m)$ is an invariant function depending on many parameters [5]. The last row in Table 4 refers to the case of the complex symmetric mass matrices and ϕ_s, ϕ_d are the new physical phases associated with the right-handed mixing matrix. The dependence of these phases is the same as in the case of the box-diagram [10]. The significance of these constraints depends very strongly on G. To illustrate this, we take the masses of the first and second generation of squarks to be equal to gluino mass and take also the same mass m_3 for the left- and right-handed sbottom. Then, the function G can be reexpressed in terms of the function F, introduced earlier in Eq. (6),

$$G(m_3, m) = F(m/m_3) - F(m_3/m) \qquad (8)$$

In Fig. 3 we plot G and G^4 as a function of the ratio m/m_3. As in the previous case, there is a significant sensitivity to the part of the parameter space where gluino and first generation of squarks are significantly heavier than m_3. The comparison with the box diagram is also very much dependent of the m/m_3 ratio. If this ratio is large $m/m_3 > 2$, the critical value of $tan\beta$ given by eq.(7) is approximately the same as in the case of completely decoupled first and second generations of squarks.

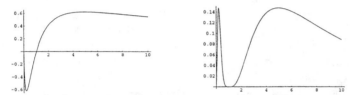

FIGURE 3. $SO(10)$ and left-right $SUSY$ models. G and G^4 are plotted as a function of m/m_3 ratio taking $m = m_\lambda = m_1 = m_2$.

IV CONCLUSIONS

We have studied in details FCNC processes mediated by Higgs particles in the supersymmetric models with large $\tan\beta$. Large $\tan\beta$ and $\mu \sim m_{sq} \sim m_\lambda$ invoke significant renormalization of M_d, mass matrix for the Down-type of quarks. It is evident that the mixing angles can also acquire additional contributions from the threshold corrections if flavour can be changed in the squark sector. When this renormalization occurs both in left- and right- handed rotation matrices, it generates flavour changing operators $\bar{d}_L b_R \bar{d}_R b_L$ and $\bar{d}_L s_R \bar{d}_R s_L$ mediated by heavy Higgs particles H and A. The coefficients in front of these operators are further enhanced by RG evolution to the low-energy scale and, in the case of neutral kaons, by chiral enhancement factor in the matrix elements.

We showed that Higgs-mediated FCNC amplitudes strongly depend on $\tan\beta$ and they grow as the fourth and sixth power of $\tan\beta$ for $\Delta B = 2$ and $\Delta S = 2$ processes respectively. As a result, Higgs-mediated amplitudes can match $SUSY$ box diagrams when $\tan\beta \sim 20$ and become the dominant mechanism for FCNC when $\tan\beta \sim m_t/m_b$. In latter case, Higgs-mediated FCNC amplitudes provide significant constraints on the off-diagonal elements in the squark mass matrices. Very strong limits on the soft-breaking sector are formulated here in Tables 1 and 2. These limits are complementary to those provided by box diagrams [3]. The significance of this mechanism varies from one model to another and depends mainly on the nontrivial structure of the right-handed squark mass matrix. For the minimal supergravity scenario this matrix is trivial and Higgs exchange mechanism is not important for any value of $\tan\beta$. Significant contributions to Δm_K, Δm_B and ϵ_K can be induced in other types of $SUSY$ models where right-handed squark mass matrices are similar to the left-handed ones.

Acknowledgements

This work is supported in part by N.S.E.R.C. of Canada. The work of A.R. is supported by an FCAR grant.

REFERENCES

1. A.E. Nelson and L. Randall, Phys. Lett. **B316** (1993) 516; R. Rattazzi and U. Sarid, Nucl. Phys. **B501** (1997) 297; T. Blazek and S. Raby, Phys.Rev.D59:095002,1999; H. Baer et al., Phys.Rev.D58:015007,1998.
2. M. Dugan, B. Grinstein and L. Hall, Nucl. Phys. **B255** (1985) 413.
3. F. Gabbiani et. al., Nucl. Phys. **B447** (1996) 321; P.H. Chankowski and S. Pokorski, hep-ph/9703442.
4. J.F. Gunion, H.E. Haber, G.L. Kane and S. Dawson, *The Higgs hunter's guide*, Addison-Wesley, 1990.
5. C. Hamzaoui, M. Pospelov and M. Toharia, Phys. Rev. D59:095005, (1999).
6. L. Wolfenstein, Comments Nucl. Part. Phys. **21** (1994) 275.
7. A.G. Cohen, D.B. Kaplan and A.E. Nelson Phys. Lett. **B388** (1996) 588; A.G. Cohen, D.B. Kaplan, F. Lepeintre, and A.E. Nelson, Phys. Rev. Lett. **78** (1997) 2300.
8. C. Hamzaoui and M. Pospelov, Euro. Phys. J. C 8, 151-156 (1999).
9. S. Dimopoulos, L.J. Hall and S. Raby, Phys. Rev. Lett. **68** (1992) 1984, Phys. Rev. **D45** (1992) 4192; P. Ramond, R.G. Roberts and G.G. Ross, Nucl. Phys. **B406** (1993) 19.
10. R. Barbieri, L.J. Hall and A. Strumia, Nucl. Phys. **B449** (1995) 437.
11. R. Kuchimanchi and R.N. Mohapatra, Phys. Rev. Lett. **75** (1995) 3989; C. S. Aulakh et al, Phys.Rev.D58:115007,1998.
12. R. Mohapatra and A. Razin, Phys. Rev. **D54** (1996) 5835.
13. M. Pospelov, Phys. Lett. **B391** (1997) 324.
14. S. Dimopoulos and L.J. Hall, Phys. Lett. **B344** (1995) 185.

Inflation from Extra Dimensions

James M. Cline

Physics Department, McGill University,
3600 University Street, Montréal, Québec, Canada H3A 2T8

Abstract. The radial mode of n extra compact dimensions (the radion, b) can cause inflation in theories where the fundamental gravity scale, M, is smaller than the Planck scale M_P. For radion potentials $V(b)$ with a simple polynomial form, to get the observed density perturbations, the energy scale of $V(b)$ must greatly exceed $M \sim 1$ TeV: $V(b)^{1/4} \equiv M_V \sim 10^{-4} M_P$. This gives a large radion mass and reheat temperature $\sim 10^9$ GeV, thus avoiding the moduli problem. Such a value of M_V can be consistent with the classical treatment if the new dimensions started sufficiently small. A new possibility is that b approaches its stable value from above during inflation. The same conclusions about M_V may hold even if inflation is driven by matter fields rather than by the radion.

INTRODUCTION

The realization that compactified extra dimensions might have radii as large as 1 mm without contradicting any current experiments [1], and that this could explain why the Higgs boson mass, m_h, is much less than the Planck scale, has created a major new direction of research in particle physics. As long as the Standard Model particles are somehow stuck on our three-dimensional subspace (brane) of the $3+n$ total spatial dimensions, the effects of the new large dimensions would have so far escaped our notice. In this picture, the Planck scale, M_P, arises as a by-product of a much smaller fundamental scale, M, and the radii of the compact dimensions, b_0, through the relation

$$M_P = 4\sqrt{\pi} M (M b_0)^{n/2}. \tag{1}$$

If M is at the TeV scale, then the problem of why m_h is much smaller than M_P is superseded by explaining why b_0 is much larger than M^{-1}. With large enough n this is a less severe tuning problem than the original hierarchy problem.

The major new effects for phenomenology come from gravitons and their Kaluza-Klein excitations, which by (n+4)-dimensional general covariance must be allowed to propagate in the extra dimensions, known as the bulk. Because their couplings to Standard Model particles are suppressed by M_P, these effects can be kept duly small, yet near the threshold for observation. In addition to the gravitons, there is a field, $b(t)$, associated with the variable size of the compact dimensions, called the radion. It can be thought of as the scale factor for the compact dimensions, where the spacetime metric has the Friedmann-Robertson-Walker-like form $g_{\mu\nu} = \text{diag}(-1, a^2, a^2, a^2, b^2, \ldots, b^2)$, provided that the compact space and the brane are homogeneous and isotropic. At late times, $b(t)$ must approach its asymptotic value b_0.

When considering inflation in such theories, it is necessary to allow for the evolution of $b(t)$ as well as the usual scale factor $a(t)$ which describes the growth of the large dimensions. In fact, the radion can economically play the role of the inflaton. Ref. [2] has explored some of the possibilities, giving a picture in which $b \ll b_0$ at the onset of inflation, grows slowly during inflation, and attains its ultimate size only long after the end of inflation. The kind of radion potential which would give this kind of behavior is rather complicated: for $b \sim b_i \ll b_0$, where b_i is the value of b during inflation, $V(b) \sim b^{2n}$; for $b_i \ll b \ll b_0$, $V(b) \sim b^{-p}$, with $p > 0$; and for $b \gg b_0$ one expects that $V(b) \sim b^n$, since this is how the contribution from the bulk cosmological constant scales with b. The complications arise from the requiring the right magnitude of density perturbations without having to introduce any energy scales which are much greater than the new gravity scale M.

In this work [3] we take a different attitude; we assume that the radion potential has a relatively simple form, and then ask, what are the consequences? A major one is that, in order to obtain large enough density perturbations, $V(b)$ must be much greater than $(1 \text{ TeV})^4$ during inflation. While presenting a problem of naturalness, this condition at the same time solves the moduli problem of the radion. In this scenario the radion stabilizes immediately following inflation. There is also a new possibility: b can overshoot its stable value and approach b_0 from above during inflation. We will present numerical solutions of the evolution equations to illustrate these outcomes for specific choices of the radion potential. We also consider conventional inflation driven by matter fields after the radion has stabilized. Interestingly, a similar restriction on $V(b)$ arises as in the case when the radion is the inflaton, to avoid runaway inflation of the compact dimensions.

EVOLUTION OF SCALE FACTORS

We start by recapitulating the equations of motion and the consistency condition for a and b, derived in [2,4]. These can be expressed in terms of Hubble parameters $H_a = \dot{a}/a$ and $H_b = \dot{b}/b$,

$$\frac{\ddot{a}}{a} + 2H_a^2 + nH_aH_b = \frac{C_n}{b^n}\left(\left(b\frac{d}{db} - (n-2)\right)V + S_a\right)$$

$$\frac{\ddot{b}}{b} + (n-1)H_b^2 + 3H_aH_b = \frac{C_n}{b^n}\left(\left(4 - 2\frac{b}{n}\frac{d}{db}\right)V + S_b\right)$$

$$3H_a^2 + \frac{1}{2}n(n-1)H_b^2 + 3nH_aH_b = (n+2)\frac{C_n}{b^n}(V + \rho) \qquad (2)$$

where $C_n = (2(n+2)M^{n+2})^{-1}$, and $V(b)$ is a potential whose minimum b_0 must be consistent with eq. (1), and such that $V(b_0) \cong 0$ so that there is no cosmological constant at the end of inflation. If there is significant pressure and energy density from matter fields, say a scalar ϕ with potential V_ϕ, then $S_a = \rho + (n-1)p$, $S_b = \rho - 3p$, $\rho = \frac{1}{2}\dot\phi^2 + V_\phi$ and $p = \frac{1}{2}\dot\phi^2 - V_\phi$, with the equation of motion $\ddot\phi + 3H_a\dot\phi + V'_\phi = 0$. We shall at first assume that $S_{a,b}$ are negligible compared to $V(b)$, however.

To get inflation of a from V, it is necessary that b starts out away from its minimum, $b_i \neq b_0$, and rolls slowly toward b_0. It is clear from eqs. (2) that the slow-roll condition is $(4 - 2\frac{b}{n}\frac{d}{db})V \ll (b\frac{d}{db} - n + 2)V$ [2]. This is only satisfied for a range of b if V has the leading behavior $V(b) \sim b^{2n}$ during inflation. For this work we will consider a potential of the form

$$V(b) = M_V^4 \left(\hat b^\alpha - \left(\frac{\alpha - \gamma}{\beta - \gamma}\right) \hat b^\beta + \left(\frac{\alpha - \beta}{\beta - \gamma}\right) \hat b^\gamma \right) \qquad (3)$$

where $\hat b = b/b_0$. It has the required properties that $V = V' = 0$ at $b = b_0$; it is thus the simplest viable form. There are two possibilities for getting inflation: either $b_i < b_0$, and $\alpha = 2n < \beta < \gamma$, or $b_i > b_0$ and $\alpha = 2n > \beta > \gamma$. The shape of V in the two cases is illustrated in figure 1. The initial condition where $b_i < b_0$ was investigated in ref. [2]; here we will consider both possibilities. Notice that in either case, the middle term in V is what drives b during inflation, since $(2 - \frac{b}{n}\frac{d}{db})$ must annihilate the dominant b^{2n} term in V in order for b to roll slowly.

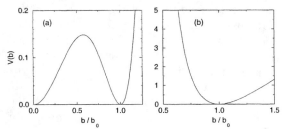

Figure 1: radion potential for the case when (a) $b_i < b_0$, so $V \sim b^{2n}$ at small b, and (b) $b_i > b_0$, so $V \sim b^{2n}$ at large b.

To numerically investigate the evolution of the scale factors and the density perturbations produced during inflation, and also for analytical insight, it is sometimes helpful to go to a dimensionless, conformal-like time variable τ, defined by $db/dt = f(b)\, db/d\tau$. During inflation, it is convenient to choose $f^2 = C_n b^{-n}(b\frac{d}{db} - n + 2)V \approx (n+2)C_n b^{-n}V$, while letting f become constant at the end of inflation. Then the scale factor has the simple time dependence $a(t) \cong e^{\tau/\sqrt{3}}$ during inflation.

The amount of inflation is determined mainly by b_i/b_0, the ratio of b's initial and final values. We will first discuss the case where $b_i < b_0$. By approximately solving the radion equation of motion when $\dot b \ll H_b \ll H_a$, one finds the conformal time dependence of b to be

$$\tau = C_b \left((b_0/b_i)^{\beta-2n} - (b_0/b)^{\beta-2n} \right) \tag{4}$$

with $C_b = [\sqrt{3}n(n+2)(\gamma-\beta)]/[2(\gamma-2n)(\beta-2n)^2]$. Thus the duration of inflation in conformal time goes like $(b_0/b_i)^{\beta-2n}$. Typically the conservative minimum of ~ 70 e-foldings of inflation (hence $\Delta\tau \approx 120$) can be achieved starting with modest initial conditions, e.g., $b_i/b_0 \leq 0.07$, as illustrated in figure 2.

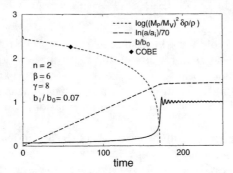

Figure 2: $\log_{10}((M_P/M_V)^2 \delta\rho/\rho)$, $\ln(a(t)/a_i)/70$ and $b(t)/b_0$, and as a function of conformal time, as described in the text. The epoch at which COBE fluctuations were produced is marked by the diamond.

DENSITY PERTURBATIONS AND REHEATING

In addition to getting enough inflation, we must also insure that the observed magnitude of density perturbations is generated. At COBE (Cosmic Background Explorer) scales, $\delta\rho/\rho \approx 1 \times 10^{-5}$. In the present theory, it is straightforward to show that [2]

$$\frac{\delta\rho}{\rho} = \frac{5}{12\pi\sqrt{2n(n-1)}} \frac{H_a^2}{(bM)^{n/2} M H_b}, \tag{5}$$

by relating the radion, b, to a canonically normalized inflaton field. We can derive a simple, accurate approximation for $\delta\rho/\rho$ during inflation by solving for H_a and H_b during the slow-roll regime, when $\ddot{b}/b \ll H_b^2 \ll \ddot{a}/b \cong H_a^2$. Recalling that $V(b) \sim M_V^4$, one obtains, at an epoch when $b = b_*$,

$$\delta\rho/\rho(b_*) = C_\rho (M_V/M_P)^2 (b_0/b_*)^{\beta-2n}, \tag{6}$$

where $C_\rho = [5\sqrt{n}(n+2)(\beta-\gamma)]/[3\sqrt{3(n-1)}(2n-\beta)(2n-\gamma)]$. The most natural value for M_V is the gravity scale M. If M is only 1 TeV, as desired for solving the Higgs mass hierarchy problem, the factor $(M/M_P)^2 \sim 10^{-32}$ suppresses $\delta\rho/\rho$ enormously. To compensate, one might be tempted to take the initial value of b/b_0 to be of order $b_i/b_0 \sim (100\, M_V/M_P)^{2/(\beta-2n)}$.

However, very small values of b_i cause inflation to last much longer, and the perturbations with large $\delta\rho/\rho$ were produced so early that they are still beyond the present horizon. COBE perturbations, on the other hand, had a wavelength of $\lambda_d = 7 \times 10^5$ ly at the time of photon decoupling (when $T = T_d = 0.25$ eV), and were produced N e-foldings before the end of inflation, with N given by

$$N \approx \ln\left(\lambda_d H_a \frac{T_d}{T_{rh}}\right) \sim \left(24 - \frac{14n}{\beta - 2n}\right) \ln 10. \tag{7}$$

The second equality was obtained by assuming $b_i/b_0 \sim (100 \, M_V/M_P)^{2/(\beta-2n)}$, $M_V = M = 1$ TeV, and a reheat temperature of 100 GeV; it shows that N is at most of order 10 under these assumptions. Thus the relevant perturbations were produced near the end of inflation. They are not enhanced by making b_i smaller; their size is determined by the value $b = b_*$ near the end of inflation, which is relatively close to b_0.

From the above results, the only way to get a large enough $\delta\rho/\rho$ with our radion potential is to give it an energy scale $M_V \cong 3 \times 10^{15}$ Gev, which is much greater than the desired low gravity scale. It is difficult to explain why M_V should be so large if $M \sim 1$ TeV. On the other hand, there must be a large mass scale hidden in $V(b)$ in any case to obtain $b_0 \gg M^{-1}$. Moreover, a large scale for $V(b)$ solves a second pernicious problem: that of reheating [1,6]. The radion mass goes like $m_b \sim M_V^2/M_P \sim 10^{12}$ GeV, so it is no longer a modulus which oscillates practically forever without decaying, as in the case when $M_V = 1$ TeV. It decays quickly ($\Gamma_b \sim g_* m_b^3/M_P^2$) into all g_* species of lighter particles, since it couples to the trace of the stress-energy tensor.

The conventional theory of reheating then gives a reheat temperature of $T_{rh} \sim 10^{-1}\sqrt{\Gamma_b M_P} \sim 10^9$ GeV [7] from the decay of the radion condensate. (We verified that the resulting energy density is less than $V(b)$ at the end of inflation.) Although such a large T_{rh} will bring bulk gravitons into thermal equilibrium, they decay with a rate $\Gamma \sim g_* T_{rh}^3/M_P^2$, thus disappearing well before nucleosynthesis [8], and even before baryogenesis; they decay in equilibrium. Although some gravitons are produced with lower energy E, hence longer lifetimes, their numbers are suppressed by the factor $(E/T_{rh})^n$ coming from the production cross section. Those with $E < 10$ TeV, which would survive to the era of nucleosynthesis and destroy light elements during their decays, are therefore diluted by a factor of $\sim 10^{5n}$ relative to photons.

Once the reheating temperature is known, eq. (7) can be evaluated again to show that the COBE fluctuations were actually produced at $N = 73 + n\ln(b_*/b_0)$ e-foldings before inflation. This implicitly determines the value of b_* since $N = (\Delta\tau - \tau_*)/\sqrt{3}$, where $\Delta\tau$ is the duration of inflation, and b depends on τ according to eq. (4). Numerically solving in the example of figure 2 shows that the COBE perturbations were produced at $\tau_* = 60$, $b_*/b_0 = 0.09$. The perturbations are rather flat in this region, with a spectral index of $n_\rho - 1 = 2d\ln(\delta\rho/\rho)/d\ln(a) = -0.015$. This is well within the present observational limits, $|n_\rho - 1| < 0.2$.

NATURAL INITIAL CONDITIONS

One of the difficulties of making $M_V \gg M$ is the validity of the low-energy effective action which gives the equations of motion (2) [2]. One expects that regions of homogeneity, where the spatial dependence of the metric can be neglected, will have an extent of only M_V^{-1}. If this is much smaller than the size of the compact dimensions, then the assumption of homogeneity within the compact space is unjustified. However with the potential (3), it is possible to start with $V(b_i) \ll M_V^4$ since b_i might be much smaller than b_0. In such a region it is natural for the spatial fluctuations in the geometry to have a longer wavelength than b_i. The condition for having $V(b_i)^{1/4} < b_i^{-1}$ can be written as

$$b_i/b_0 < (M_P/M_V)^{2/(n+2)} (M/M_P)^{2/n}. \qquad (8)$$

If $M = 1$ TeV, then the least fine-tuned case of $n = 7$, assuming an upper limit of 11 dimensions, gives $b_i/b_0 < 10^{-4}$. In the spirit of chaotic inflation, this is not unnaturally small. As long as there are some regions in the initial universe satisfying (8), they will start to inflate, and continue for many ($\sim 10^{4(\beta - 2n)}$) e-foldings. If other regions fail to inflate because of their inhomogeneities, this need not concern us. Moreover, if M is of order M_V, eq. (8) becomes $b_i/b_0 \lesssim 10^{-12/(n(n+2))}$, which needs no fine-tuning at all if $n > 3$.

The above discussion was for $b_i < b_0$. What happens if the compact space starts out larger than its stable value? As long as the exponent β is less than $2n$ in the potential (3) and $\gamma < \beta$, the story is quite similar to the case of $b_i < b_0$. One can obtain enough inflation if $b_i/b_0 \gtrsim 6$, and the estimate of the density fluctuations again requires M_V to be near the GUT scale. However the initial conditions now appear to be quite fine-tuned: the length scale provided by the potential, M_V^{-1}, is many orders of magnitude smaller than the size of the compact space, so one must wonder how the latter came to be so smooth at the beginning of inflation. In fact, it is possible to start with $b \ll b_0$, as before, but the radion picks up so much speed that it overshoots b_0 and attains a large value which marks the onset of inflation. This situation is illustrated in figure 3. The spectral index is again $n_\rho - 1 = -0.015$ in this example.

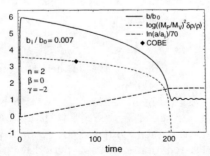

Figure 3: $b(t)/b_0$, $\log_{10}((M_P/M_V)^2 \delta\rho/\rho)$ and $\ln(a(t)/a_i)/70$ in an example where inflation occurs while $b > b_0$.

INFLATION WITH MATTER FIELDS

Unfortunately the most natural form for the radion potential is $V(b) \sim Ab^n - Bb^{n-2} + Cb^{n-2p}$ [4,5], which does not give inflation. For this reason it may be more convincing to drive inflation with conventional matter fields [9] or extra branes [10] after the radion has stabilized. An interesting implication of the extra dimensions is that they can be destabilized by a conventional inflaton, and begin inflating themselves. As one might expect, whether this occurs depends upon how stiff the radion potential is compared to the size of the perturbing matter potential, which we shall refer to as V_ϕ. Thus one obtains a constraint on $V(b)$ from the properties of the inflaton potential. (A different argument leading to a similar constraint was given in refs. [8] and [11].)

To see how such a constraint comes about, parametrize the radion by $b = b_0(1+\epsilon)$ and suppose that the potential has the form $V \sim M_V^4 \epsilon^2$ in the vicinity of b_0. Consider an inflationary phase where the pressure and energy density are dominated by the potential V_ϕ of a matter field ϕ that lives on the brane. (The same conclusions also follow if ϕ inhabits the bulk.) The source terms in eq. (2) are $S_a = (2-n)V_\phi$ and $S_b = 4V_\phi$. A nonzero S_b will induce a shift in the radion as it tries to minimize the full source term for the b equation of motion,

$$-n^{-1} M_V^4 \epsilon \left(1 - (n-1)\epsilon\right) + V_\phi \cong 0. \tag{9}$$

If $V_\phi \ll M_V^4$ we can work to linear order in ϵ. The source term for the a equation then becomes $(n+2) C_n b^{-n} V_\phi$, in accordance with the consistency condition, third of eqs. (2). However if V_ϕ becomes too large, the quadratic term in ϵ becomes important, and it is no longer possible to satisfy eq. (9). At this point b cannot resist inflation. This situation must be avoided because, once b starts inflating, it is a runaway process; b inflates forever, as shown in figure 4.

If b remains close to b_0, eqs. (2) reduce to the usual Friedmann equation, $H_a^2 = 8\pi\rho/3M_P^2$. In chaotic inflation with a potential of the form $V_\phi = M_\phi^{4-p} \phi^p$, it is straightforward to show that the relation between ϕ and a is $\phi^2 = \phi_i^2 - (pM_P^2/4\pi)\ln(a/a_i)$, so that initially $\phi_i \sim$ (several M_p) to get 70 e-foldings of a. The density perturbations during this epoch are given by $\delta\rho/\rho \sim (5/12\pi) H_a^2/\dot\phi \sim 10^3 \sqrt{V_\phi} M_P^{-2} \sim 10^{-5}$. We thus obtain

$$V_\phi \sim 10^{-16} M_P^4 \lesssim M_V^4 \tag{10}$$

which is quite similar to the bound on M_V when the radion is the inflaton. In hybrid inflation models [12] this bound is softened to $M_V > 10^{-1} M_P^{3/5} m^{2/5}$.

Finally let us ask: how problematic is it for M_V to be much larger than the gravity scale, assuming it is near 1 TeV? First, it should be kept in mind that there is still no proposal for getting the required b^{2n} behavior of $V(b)$ (nor the subsequent b^{-p} behavior in the post-inflation expansion period envisioned in [2]). Whatever mechanism that emerges to explain this might also make a large energy scale more plausible. Second, even in the most realistic model for $V(b)$, scales larger than M

may be required to explain why $b_0 \gg M^{-1}$. The term in $V(b)$ coming from $\sqrt{-g}R$ in the Einstein-Hilbert action is necessarily of order $M^4(Mb)^{n-2}$, so that when $b \sim b_0$, $V \sim M^4(M_P/M)^{2-4/n}$, which is much larger than M^4 unless $n = 2$. Although not as severe a hierarchy problem as the present one, it is qualitatively similar. Third, the M_V hierarchy problem disappears altogether if we allow M to be the GUT scale rather than 1 TeV. While this value is less exciting for current accelerator experiments, it may be nature's choice, and it still presents new possibilities for the early universe.

I thank C. Burgess, N. Kaloper, G. Moore, R. Myers and M. Paranjape for helpful observations and criticisms.

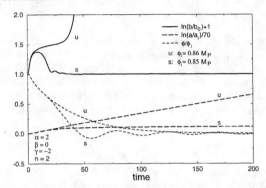

Figure 4: $\ln(b(t)/b_0) + 1$, $\ln(a(t)/a_i)/70$ and ϕ/ϕ_i for initial conditions that are stable (s) and unstable (u) against eternal inflation of b. V_ϕ is proportional to ϕ^2 in this example.

REFERENCES

1. N. Arkani-Hamed, S. Dimopoulos and G. Dvali, Phys. Lett. B429 (1998) 263; Phys. Rev. D59 (1999) 086004; I. Antoniadis, N. Arkani-Hamed, S. Dimopoulos and G. Dvali, Phys. Lett. B436 (1998) 257.
2. N. Arkani-Hamed, S. Dimopoulos, N. Kaloper and J. March-Russell, hep-ph/9903224 (1999).
3. J. Cline, hep-ph/9904495 (1999).
4. N. Arkani-Hamed, S. Dimopoulos and J. March-Russell, hep-th/9809124 (1998)
5. T. Banks, M. Dine and A. Nelson, hep-th/9903019 (1999).
6. C. Csaki, M. Graesser and J. Terning, hep-ph/9903319 (1999).
7. A.D. Linde, *Paricle Physics and Inflationary Cosmology*, Harwood Academic Publishers (1990).
8. K. Benakli and S. Davidson, hep-ph/9810280 (1998).
9. N. Kaloper and A. Linde, hep-th/9811141 (1998).
10. G. Dvali and S.H.H. Tye, Phys. Lett. B450 (1999) 72.
11. D. Lyth, Phys. Lett. B448 (1999) 191.
12. A.D. Linde, Phys. Rev. D49 (1994) 748.

Large-N Yang–Mills Theory as Classical Mechanics

C.-W. H. Lee[1] and S. G. Rajeev

Department of Physics and Astronomy, University of Rochester, P.O. Box 270171, Rochester, New York 14627.

Abstract. To formulate two-dimensional Yang–Mills theory with adjoint matter fields in the large-N limit as classical mechanics, we derive a Poisson algebra for the color-invariant observables involving adjoint matter fields. We showed rigorously in Ref. [15] that different quantum orderings of the observables produce essentially the same Poisson algebra. Here we explain, in a less precise but more pedagogical manner, the crucial topological graphical observations underlying the formal proof.

One major unsolved problem in physics is hadronic structure. We would like to explain, for instance, the momentum distributions of valence quarks, sea quarks and gluons inside a proton. We have accumulated a fairly large amount of experimental data on the distribution functions [1], but we have made relatively little advance in explaining them from the widely accepted fundamental theory of strong interaction, quantum chromodynamics (QCD).

The emergence of hadrons is a low-energy phenomenon of strong interaction. The strong coupling constant is large, and perturbative QCD fails very badly. Other approximations are needed. One widely studied approximation is the large-N limit, in which the number of colors N is taken to be infinitely large [2]. This approximation is believed to capture the essence of low-energy strong interaction phenomena. Indeed, 't Hooft showed that if space–time is assumed to be two-dimensional only, then the meson spectrum displays itself as a Regge trajectory [3]. This attractive feature of the large-N limit has drawn the attention of a large number of researchers for more than two decades. They want to build up a systematic theory of the large-N limit to deal with baryons in addition to mesons in four-dimensional space–time.

One important feature of the large-N limit is that the expectation value of a product of two observables A and B is the same as the product of the expectation values of these two observables. The difference is of order $1/N$ and so can be

---

[1] speaker.

omitted [4,5]:
$$\langle \hat{A}\hat{B}\rangle = \langle \hat{A}\rangle\langle \hat{B}\rangle + O(1/N).$$

In other words, there is no quantum fluctuation. The theory thus behaves like *classical mechanics* [6], and it should be possible for us to formulate the large-N limit of QCD as classical mechanics.

Formally speaking, we need three ingredients to build up a theory as classical mechanics [7]. The first is the notion of a manifold to describe the geometry of the phase space of positions and momenta. Dynamical variables are then functions on the manifold. The second is the notion of a Hamiltonian function H, one of the dynamical variables. This function displays the physical features (e.g., symmetry) of the system, and governs the time evolution of it. How the Hamiltonian function governs the time evolution is determined by a *Poisson algebra* [8], the third ingredient of classical mechanics. To understand what a Poisson algebra is, we need a number of preliminary notions.

An *algebra* R is a linear space on a field K with a multiplication rule of any 2 vectors in R such that for any x, y and $z \in R$ and $a \in K$,

1. $xy \in R$;

2. the vectors satisfy the distributive properties
$$x(y+z) = xy + xz \text{ and}$$
$$(x+y)z = xz + yz; \text{ and}$$

3. $a(xy) = (ax)y = x(ay)$.

An *associative algebra* is an algebra R such that for any x, y and $z \in R$,
$$x(yz) = (xy)z.$$

A *Poisson algebra* is an associative algebra R which is equipped with a blinear map $\{\,,\,\}: R \times R \to R$, the Poisson bracket, with the following properties for any x, y and $z \in R$:

1. skew-symmetry:
$$\{y, x\} = -\{x, y\};$$

2. the Jacobi identity:
$$\{x, \{y, z\}\} + \{y, \{z, x\}\} + \{z, \{x, y\}\} = 0; \text{ and}$$

3. the Leibniz identity:
$$\{xy, z\} = x\{y, z\} + \{x, z\}y.$$

How a dynamical variable $G(t)$ changes with the time t is given by the equation

$$\frac{dG(t)}{dt} = \{G.H\}.$$

Making the assumption that space–time is two-dimensional, one of us introduced a formulation of large-N Yang–Mills theory as classical mechanics several years ago [9]. As in the mesonic model 't Hooft studied, gluons are not dynamical objects in this classcial mechanics; only quarks and anti-quarks are. The phase space turns out to an infinite-dimensional Grassmannian manifold [10]. (Briefly speaking, an infinite-dimensional Grassmannian manifold is a collection of subspaces of an infinite-dimensional vector space. A differential structure is conferred upon this collection to turn it into a manifold.) Dynamical variables are composed of quark and anti-quark fields. They are bilocal functions on the Grassmannian manifold. The Poisson bracket can be uniquely determined from the geometric properties of the Grassmanian. The Hamiltonian is chosen in such a way that upon quantization of this classical mechanics, i.e., if we retain terms of subleading orders in the large-N limit, it will take exactly the form the action of Yang–Mills theory is conventionally written. The meson spectrum predicted by this classical mechanics is precisely the same as that obtained by 't Hooft. However, the model can be used to calculate structure functions of a baryon also [11]. As an initial attempt, the proton is assumed to be made up of valence quarks only. Sea quarks and gluons are omitted. In the infinite momentum frame, the transverse momenta of valence quarks can be neglected. Thus the Yang–Mills theory can be dimensionally reduced from four dimensions to two. Dimensional reduction is a good approximation for valence quarks carrying a large fraction of the total momentum of the proton, but not so good for the quarks carrying almost no momentum. Indeed, it turned out the momentum distribution functions predicted by this classical model of Yang–Mills theory agrees well with experimental data at high-momentum regime, but not so well near zero momentum.

How about sea quarks and gluons inside a hadron? The sea quark distribution function can actually be predicted within the same classical mechanical framework [9,12]; it is a matter of modifying the details of some approximations. However, gluon dynamics require a new set Poisson brackets and a new phase space. Since gluons carry about 20% of the total momentum of a proton [1], it is worthwhile to study them further.

A Poisson algebra for gluons was constructed a few years ago [13]. This was achieved by a technique called deformation quantization [14,8]. Deformation quantization refers to the procedure of defining an algebra of smooth functions in such a way that when the functions are multiplied, it is as if we are multiplying suitably ordered operators these smooth functions represent. Any physical observable of gluons must involve creation and annihilation operators of gluons, each of which carries two color quantum numbers and a vector-valued linear momentum. The two color quantum numbers can be treated as the row and column indices of an $N \times N$ matrix, N being the total number of colors. Any physical observable has to be color-invariant. This implies that it has to be a polynomial of the traces of matrix products with a generic form

$$f^I = \frac{1}{N^{m/2+1}} \mathrm{Tr}\, \eta^{i_1} \eta^{i_2} \cdots \eta^{i_m},$$

where $i_1, i_2, \ldots, i_m$ are quantum states other than colors of the gluons, I is the sequence $i_1, i_2, \ldots i_m$, and the factor of N is put to the left of the trace to ensure that the Poisson algebra of these f^I's, to be introduced shortly, is well defined in the large-N limit. In each product, the operators have to satisfy a certain ordering. As an initial attempt, the operators are Weyl-ordered in Ref. [13]. Hence,

$$\eta^i = 1/2(a_\sigma^{i\rho} + a_\sigma^{\dagger i\rho})$$
$$\eta^{-i} = i/2(a_\sigma^{i\rho} - a_\sigma^{\dagger i\rho})$$

When we multiply two physical observables together, we need to rearrange the order of the creation and annihilation operators to make the resultant product consistent with the quantum ordering. As a result, multiplication of physical observables is still associative but no longer commutative. The commutator of two physical observables therefore provides us a Poisson bracket. In the case of Weyl-ordered quantum observables of gluons, the Poisson bracket is

$$\{f^I, f^J\}_W = 2i \sum_{\substack{r=1,\mathrm{odd}}}^{\infty} \sum_{\substack{\mu_1 < \mu_2 < \cdots < \mu_r \\ (\nu_1 > \nu_2 > \cdots > \nu_r)}} \left(-\frac{i\hbar}{2}\right)^r \tilde{\omega}^{i_{\mu_1} j_{\nu_1}} \cdots \tilde{\omega}^{i_{\mu_r} j_{\nu_r}}$$
$$\cdot f^{I(\mu_1,\mu_2)J(\nu_2,\nu_1)} f^{I(\mu_2,\mu_3)J(\nu_3,\nu_2)} \cdots f^{I(\mu_r,\mu_1)J(\nu_1,\nu_r)}. \quad (1)$$

In this equation, $\tilde{\omega}^{ij}$ is an anti-symmetric constant tensor. $I(\mu_1, \mu_2)J(\nu_2, \nu_1)$ is the sequence $i_{\mu_1+1}, i_{\mu_1+2}, \ldots, i_{\mu_2-1}, j_{\nu_2+1}, j_{\nu_2+2}, \ldots, j_{\nu_1-1}$. Notice that we sum over all possible sets of values of $\mu_1, \mu_2, \ldots, \mu_r$ such that they are strictly increasing, and all possible sets of values of $\nu_1, \nu_2, \ldots, \nu_r$ such that they are strictly decreasing up to a cyclic permutation. Eq.(1) looks complicated, but we can actually visualize it in Fig. 1. We will call this Poisson algebra $\mathcal{W}$.

Nevertheless, in many practical applications, we need quantum observables made up of normal-ordered rather than Weyl-ordered operators. We thus need to apply the above idea of deformation quantization to derive a Poisson bracket of normal-ordered observables [15]. In this case, A color-invariant observable has the form

$$\phi^I = \frac{1}{N^{n/2+1}} \mathrm{Tr}\, z^{i_1} z^{i_2} \cdots z^{i_n},$$

where

$$z_\sigma^{i\rho} = a_\sigma^{i\rho} \text{ and } z_\sigma^{-i\rho} = a_\sigma^{\dagger i\rho}$$

for $i > 0$. The Poisson bracket for these operators turns out to be

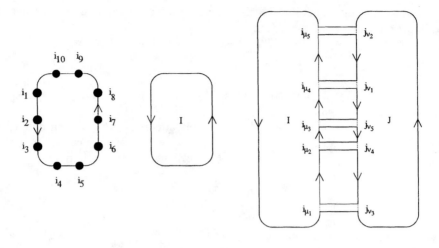

FIGURE 1. (a) A typical color-invariant observable f^I. Each solid circle represents an η^i. Notice the cyclic symmetry of the figure. (b) A simplified diagrammatic representation of f^I. We use the capital letter I to denote the whole sequence i_1, i_2, ..., i_m. (c) A typical term in $\{f^I, f^J\}_W$. This is a product of the color-invariant observables $f^{I(\mu_1,\mu_2)J(\nu_4,\nu_3)}$, $f^{I(\mu_2,\mu_3)J(\nu_5,\nu_4)}$, $f^{I(\mu_3,\mu_4)J(\nu_1,\nu_5)}$, $f^{I(\mu_4,\mu_5)J(\nu_2,\nu_1)}$, and $f^{I(\mu_5,\mu_1)J(\nu_3,\nu_2)}$. We can identify these color-invariant observables by their vertices. For example, $f^{I(\mu_1,\mu_2)J(\nu_4,\nu_3)}$ can be described as a 'loop with vertices i_{μ_1}, i_{μ_2}, j_{ν_4} and j_{ν_3}', though none of these vertices belong to $f^{I(\mu_1,\mu_2)J(\nu_4,\nu_3)}$.

$$\{\phi^I, \phi^J\}_N = \sum_{r=1}^{\infty} \sum_{\substack{\mu_1 < \mu_2 < \ldots < \mu_r \\ (\nu_1 > \nu_2 > \ldots > \nu_r)}} \hbar^r \gamma^{i_{\mu_1} j_{\nu_1}} \ldots \gamma^{i_{\mu_r} j_{\nu_r}}$$
$$\cdot \phi^{I(\mu_1,\mu_2)J(\nu_2,\nu_1)} \phi^{I(\mu_2,\mu_3)J(\nu_3,\nu_2)} \ldots \phi^{I(\mu_r,\mu_1)J(\nu_1,\nu_r)} - (I \leftrightarrow J), \qquad (2)$$

where $\gamma^\mu_\nu = 0$ unless $\mu < 0$ and $\nu > 0$. We will call this Poisson algebra $\mathcal{N}$.

The Poisson algebras $\mathcal{W}$ and $\mathcal{N}$ look very different. Are they intrinsically different Poisson algebras, or are they the same Poisson algebra with different expressions? This question can be answered by looking for a Poisson morphism between $\mathcal{W}$ and $\mathcal{N}$. Let R_1 and R_2 be two Poisson algebras. A *Poisson morphism* is a mapping $F : R_1 \to R_2$ such that it preserves

1. vector addition:
$$F(x+y) = F(x) + F(y);$$

2. scalar multiplication:
$$F(kx) = kF(x);$$

3. vector multiplication:
$$F(xy) = F(x)F(y); \text{ and}$$

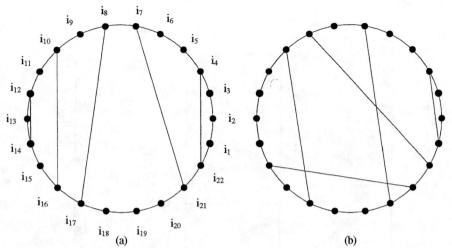

FIGURE 2. (a) An allowable partition of a color-invariant observable into a product of color-invariant observables. Note that no two lines cross each other. (b) A forbidden partition of the color-invariant observable. Note that some lines cross one another.

4. the Poisson bracket:
$$F(\{x,y\}_1) = \{F(x), F(y)\}_2$$

for any $k \in K$, x and $y \in R_1$. Here $\{\,,\,\}_1$ and $\{\,,\,\}_2$ are the Poisson brackets of R_1 and R_2, respectively. If there exists a Poisson morphism between two Poisson algebras, then they are effectively the same Poisson algebra.

In our case, a Poisson morphism $F: \mathcal{W} \to \mathcal{N}$ *does exist*. The reader can find a thoroughly rigorous formulation and proof of this Poisson morphism in Ref. [15]. Roughly speaking, the mapping F is accomplished in two steps. Consider the color-invariant observable $f^{i_1 \cdots i_{22}} \in \mathcal{W}$ in Fig. 2. The first step involves splitting the big all-encompassing loop into a number of smaller loops. The straight lines joining the solid circles and cutting off the big loop have to be within the loop. Moreover, no two straight lines can cross each other. Each solid circle η^i which is not touched by any straight line is now identified as a linear combination of solid circles η'^i in $\mathcal{N}$ by the formula

$$\eta'^i = \begin{cases} \frac{1}{2}(z^i + z^{-i}) & \text{if } i > 0; \text{ and} \\ \frac{1}{2}(z^i - z^{-i}) & \text{if } i < 0. \end{cases}$$

The resultant diagram represents the product of the color-invariant observables in $\mathcal{N}$, each of which is represented by a smaller loop in the resultant diagram. For instance, the product in Fig. 2(a) is

$$T^{i_4 i_{22}} T^{i_7 i_{21}} T^{i_8 i_{17}} T^{i_{10} i_{16}} T^{i_{12} i_{14}} \phi^{i_1 i_2 i_3} \phi^{i_5 i_6} \phi^{i_9} \phi^{i_{11} i_{15}} \phi^{i_{13}} \phi^{i_{18} i_{19} i_{20}},$$

where T^{ij} is a polynomial of C^{ij} and C^{ji}, where in turn C^{ij} is a polynomial of γ^{ij},

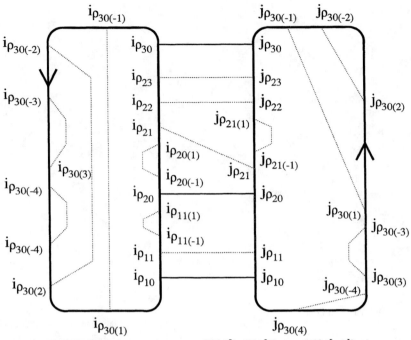

FIGURE 3. A typical term in $\{F(f^I), F(f^J)\}_N$ or $F(\{f^I, f^J\}_W)$.

$\gamma^{-i,j}$, $\gamma^{i,-j}$ and $\gamma^{-i,-j}$ [15]. The second step involves summing over all allowable partitions of f^I in $\mathcal{W}$.

Why is this F a Poisson morphism? The most non-trivial statement we need to show is that F preserves the Poisson bracket. In other words, We need to show that each term in $\{F(f^I), F(f^J)\}_N$ is a term in $F(\{f^I, f^J\}_W)$, and vice versa. Consider Fig. 3. The big oval-shaped object on the left is f^I, and the one on the right is f^J. The dotted lines inside f^I divide it into a product of color-invariant observables in $\mathcal{N}$. So do the dotted lines insider f^J. A typical term in $\{F(f^I), F(f^J)\}_N$ is obtained by choosing a smaller loop in I, and a smaller loop in J, and perform the contractions illustrated in Fig. 1. In Fig. 3, we have chosen the smaller loop with vertices $i_{\rho_{11(-1)}}$, $i_{\rho_{11(1)}}$, $i_{\rho_{20(-1)}}$, $i_{\rho_{20(1)}}$, $i_{\rho_{30(-1)}}$ and $i_{\rho_{30(1)}}$ in I, and the smaller loop $j_{\rho_{21(-1)}}$, $j_{\rho_{21(1)}}$, $j_{\rho_{30(-1)}}$, $j_{\rho_{30(1)}}$, $j_{\rho_{30(-3)}}$, $j_{\rho_{30(3)}}$, $j_{\rho_{30(-4)}}$ and $j_{\rho_{30(4)}}$ in J. (See the explanation of the jargons here in the caption of Fig. 1.) There are 7 contractions due to the Poisson bracket. The contracted pairs are $i_{\rho_{10}}$ and $j_{\rho_{10}}$, $i_{\rho_{11}}$ and $j_{\rho_{11}}$, $i_{\rho_{20}}$ and $j_{\rho_{20}}$, $i_{\rho_{21}}$ and $j_{\rho_{21}}$, $i_{\rho_{22}}$ and $j_{\rho_{22}}$, $i_{\rho_{23}}$ and $j_{\rho_{23}}$, and $i_{\rho_{30}}$ and $j_{\rho_{30}}$. According to Eq.(1), these contractions produce in this term of $\{F(f^I), F(f^J)\}_N$ a constant factor G which the reader can divine is a polynomial of $C^{\pm i_{\rho_{10}}, \pm j_{\rho_{10}}}, \ldots, C^{\pm i_{\rho_{30}}, \pm j_{\rho_{30}}}$. Now notice a crucial observation. Identify $\tilde{\omega}^{ij}$ with a certain polynomial of C^{ij} and C^{ji}. Replace an *odd* number of C^{ij}'s in G with $\tilde{\omega}^{ij}$'s, and the remaining C^{ij}'s with T^{ij}'s. Add up all such possible replacements. The sum will precisely be G up to a multiplicative constant. Consequently, we can draw some of these contractions as dotted lines because they are T's, and others as thick lines to show that they are $\tilde{\omega}$'s. In Fig. 3, $C^{i_{\rho 10} j_{\rho 10}}$, $C^{i_{\rho 20} j_{\rho 20}}$ and $C^{i_{\rho 30} j_{\rho 30}}$ are replaced with $\tilde{\omega}^{i_{\rho 10} j_{\rho 10}}$, $\tilde{\omega}^{i_{\rho 20} j_{\rho 20}}$ and $\tilde{\omega}^{i_{\rho 30} j_{\rho 30}}$, respectively, whereas $C^{i_{\rho 11} j_{\rho 11}}$, $C^{i_{\rho 21} j_{\rho 21}}$, $C^{i_{\rho 22} j_{\rho 22}}$ and $C^{i_{\rho 23} j_{\rho 23}}$ are replaced with $T^{i_{\rho 11} j_{\rho 11}}$, $T^{i_{\rho 21} j_{\rho 21}}$, $T^{i_{\rho 22} j_{\rho 22}}$ and $T^{i_{\rho 23} j_{\rho 23}}$, respectively.

Now notice another crucial observation. This diagram can be reproduced by computing the Poisson bracket in $\mathcal{W}$ first and mapping the resultant expression with F later. The Poisson bracket in $\mathcal{W}$ produces a product of three color-invariant observables. The first one L_1 is characterized with the vertices $i_{\rho 10}$, $i_{\rho 20}$, $j_{\rho 20}$ and $j_{\rho 10}$; the second one L_2 with the vertices $i_{\rho 20}$, $i_{\rho 30}$, $j_{\rho 30}$ and $j_{\rho 20}$; and the third one L_3 with the vertices $i_{\rho 30}$, $i_{\rho 10}$, $j_{\rho 10}$ and $j_{\rho 30}$. The mapping F projects this product to $\mathcal{N}$. This is done, as usual, by splitting these 3 loops into smaller loops with dotted lines so that no two lines cross each other inside any of these 3 loops. Notice that in Fig. 3, if we flip the dotted line $i_{\rho_{11(-1)}} i_{\rho_{11(1)}}$ to within L_1, and the dotted lines $i_{\rho_{20(-1)}} i_{\rho_{20(1)}}$ and $j_{\rho_{21(-1)}} j_{\rho_{21(1)}}$ to within L_2, no two dotted lines will cross each other. Therefore, we obtain the same term.

Conversely, any term $F(\{f^I, f^J\}_W)$ can be seen as a term in $\{F(f^I), F(f^J)\}_N$ by a simiar diagrammatic argument. Thus F is indeed a Poisson morphism.

This Poisson algebra may have other interesting mathematical properties. We hope that we can use this Poisson algebra to solve gluon dynamics in the future.

We thank O. T. Turgut for discussions. This work was supported in part by the U.S. Department of Energy under grant DE-FG02-91ER40685.

REFERENCES

1. R. Brock et al., Rev. Mod. Phys. **67**, 157 (1995).
2. G. 't Hooft, Nucl. Phys. **B72**, 461 (1974).
3. G. 't Hooft, Nucl. Phys. **B75**, 461 (1974).
4. E. Witten, Nucl. Phys. **B160**, 57 (1979).
5. S. Coleman, *Aspects of Symmetry*, Cambridge University Press, Cambridge, 1985.
6. L. Yaffe, Rev. Mod. Phys. **54**, 407 (1982).
7. V. I. Arnold, *Mathematical Methods of Classical Mechanics*, 2nd ed., Springer-Verlag, New York, 1989.
8. V. Chari and A. Pressley, *A Guide to Quantum Groups*, Cambridge University Press, Cambridge, 1994.
9. S. G. Rajeev, Int. J. Mod. Phys. **A9**, 5583 (1994); hep-th/9905072.
10. A. Pressley and G. Segal, *Loop Groups*, Clarendon, Oxford, 1986.
11. G. S. Krishnaswami and S. G. Rajeev, Phys. Lett. **B441**, 429 (1998).
12. V. John, G. S. Krishnaswami and S. G. Rajeev, in preparation.
13. S. G. Rajeev and O. T. Turgut, J. Math. Phys. **37**, 637 (1996).
14. M. Flato, A. Lichnerowicz and D. Sternheimer, C. R. Acad. Sci. Paris Sér. **A279**, 877 (1974); Composito Mathematica **31**, 47 (1975); J. Math. Pure Appl. **54**, 445 (1975).
15. C.-W. H. Lee and S. G. Rajeev, J. Math. Phys. **40**, 1870 (1999).

Gauge Field Correlators in 4-D Simplicial Quantum Gravity

Eric B. Gregory and Simon M. Catterall

201 Physics Building, Syracuse University
Syracuse, NY, USA, 13210-1130

Abstract. We discuss the motivation for studying gauge fields coupled to four-dimensional simplicial quantum gravity. We describe some of the issues involved in measuring connected correlators on dynamical geometry and show possible evidence for a massless propagating mode.

I THE MODEL AND SIMULATION

Our interest is in the path integral quantization of Einstein-Hilbert gravity in four dimensions coupled to massless vector gauge fields. Gravity is notoriously stubborn in resisting analytic path integral quantization. The numerical approach of simplicial quantum gravity is a common alternative. In simplicial quantum gravity, a manifold of arbitrary geometry is approximated by the gluing together of small pieces of flat space called *simplices*. A simplex in d dimensions is a collection of $d+1$ points each connected to all the others by links of unit length. In 2 dimensions the simplex is an equilateral triangle. The three-simplex is a tetrahedron and in four dimensions the simplex is the hyper-tetrahedron. The four-simplices are glued together along their tetrahedral faces to form a discretization of a curved manifold of Euclidean signature.

An appropriate sampling of the space of all possible ways of assembling the simplices approximates the Euclidean integration over metrics:

$$Z = \int \mathcal{D}g_{\mu\nu}\mathcal{D}Ae^{-S_g[A]} \longrightarrow Z = \sum_{\tau}\sum_{\{A\}} s^{-S_\tau[A]}. \qquad (1)$$

Here we have used A to represent generic matter fields. We are in fact interested in the case of one or more vector gauge fields coupled to the geometry, that is,

$$S_g[A]\frac{1}{16\pi G}\int d^4x\sqrt{g}(2\Lambda - R) + \int d^4x\sqrt{g}F_{\mu\nu}F^{\mu\nu}. \qquad (2)$$

Under the discretization described above, equation (2) becomes

$$S_\tau[A] = \kappa_4 N_4 - \kappa_0 N_0 + \sum_{ijk \in \text{triangle}} (A_{ij} + A_{jk} + A_{ki})^2. \tag{3}$$

II BACKGROUND AND MOTIVATION

By the late 1990's it had become clear that a lattice formulation of pure Einstein-Hilbert gravity contains no continuous phase transition which is a necessity for a continuum limit to describe a non-trivial theory. Numerical studies [1] have found that the phase transition encountered when tuning κ_0 is first order. Clearly this is bad news for simplicial quantum gravity as a avenue to an ultraviolet quantum theory.

However, it may be that there is a phase where the *large-scale* physics is effectively described by an effective action cast in terms of a conformal field σ. [3] [4] [5] That is, one can approximate the long-range features of space-time using a metric that can be conformally transformed to a flat space metric $\bar{g}_{\mu\nu}$:

$$g_{\mu\nu} = e^\sigma \bar{g}_{\mu\nu}. \tag{4}$$

Unlike in the case of two-dimensional quantum gravity, this is a severe truncation of the full theory.

This conformal phase would be reachable if there were enough matter coupled to the the geometry as parameterized by

$$Q^2 = \frac{1}{180}\left(N_S + \frac{11}{2} N_{\text{WF}} + 62 N_V - 28\right) + Q^2_{\text{grav}}, \tag{5}$$

where N_S, N_{WF}, and N_V are the number of scalar fields, Weyl fermions, and vector fields coupled to the geometry and Q^2_{grav} is the (unknown) graviton contribution to Q^2. The -28 is the contribution of the conformal field σ. Antoniadis suggested in [3] that to access the conformal phase we must have $Q^2 > 4$. The parameter Q^2 is somewhat analogous to the central charge c in two-dimensional gravity. To investigate this proposed phase we must couple enough matter fields to drive Q^2 above 4. It is very clear from equation (5) that the most efficient way of driving up the value of Q^2 is to add vector fields. For example, to increase Q^2 by one we could add 180 scalar fields or three vector fields. Particularly in the case of numerical studies, then, vector fields are the way to investigate the existence of this new phase.

III MEASURING CONNECTED CORRELATORS IN SIMPLICIAL QUANTUM GRAVITY

Measuring connected correlators in quantum gravity requires more sophistication than the same procedure in flat space. In flat space we measure

$$\langle QQ \rangle_c(R) = \langle [Q_{\vec{m}} - \langle Q \rangle][Q_{\vec{m}+\vec{r}} - \langle Q \rangle] \rangle_{m,|\vec{r}|=R}$$
$$= \langle Q_{\vec{m}} Q_{\vec{m}+\vec{r}} - Q_{\vec{m}} \langle Q \rangle - \langle Q \rangle Q_{\vec{m}+\vec{r}} + \langle Q \rangle^2 \rangle_{m,|\vec{r}|=R}$$
$$= \langle Q_{\vec{m}} Q_{\vec{m}+\vec{r}} - \langle Q \rangle^2 \rangle_{m,|\vec{r}|=R}. \qquad (6)$$

Let's instead be more careful and write this as

$$\langle QQ \rangle_c(R) = \frac{1}{V} \frac{1}{n(R)} \langle \sum_m \sum_n [(Q_m - \langle Q \rangle)(Q_n - \langle Q \rangle)] \delta_{D_{mn},R} \rangle. \qquad (7)$$

Here V is the number of sites in the lattice, and $n(R)$ is the number of lattice sites that are a distance R away from an initial site (more on this in a moment). Finally the delta function allows the contribution only of those terms where the the distance D_{mn} between sites m and n is R.

We can reconcile expressions (6) and (7) only because:

- We have a definition of distance $D_{mn} = |\vec{m} - \vec{n}|$
- The geometry and field configurations are independent.

Things are slightly more involved in the case of fluctuating geometry. First let us define D_{mn} as the geodesic distance between simplices m and n (i.e. dual lattice sites). In other words D_{mn} is the minimum number of simplex-to-simplex hops that can get us from simplex m to simplex n. Keeping this in mind, we can now evaluate (6)

$$\langle QQ \rangle_c(R) = \frac{1}{V} \frac{1}{n(R)} \langle \sum_m \sum_n Q_m Q_n \delta_{D_{mn},R}$$
$$-2\langle Q \rangle \sum_m Q_m \sum_n \delta_{D_{mn},R}$$
$$+\langle Q \rangle^2 \sum_m \sum_n \delta_{D_{mn},R} \rangle. \qquad (8)$$

Note that only in the case where the geometry is not coupled to the fields can we say that

$$\frac{1}{V} \frac{1}{n(R)} \sum_m Q_m \sum_n \delta_{D_{mn},R} = \frac{1}{V} \frac{1}{n(R)} \left[\sum_m Q_m \right] [n(R)] = \langle Q \rangle. \qquad (9)$$

Otherwise, as is in the case of quantum gravity, we must measure three different quantities [2]:

$$G^{QQ}(R) = \frac{1}{V} \sum_m \sum_n Q_m Q_n \delta_{D_{mn},R} \qquad (10)$$

$$G^{Q1}(R) = \frac{1}{V} \sum_m Q_m \sum_n \delta_{D_{mn},R} \qquad (11)$$

$$n(R) = \frac{1}{V} \sum_m \sum_n \delta_{D_{mn},R}. \tag{12}$$

On fluctuating geometry, none of these quantities are static and so we must measure these on each configuration and take expectation values. Then we can say

$$\langle QQ \rangle_c(R) = \frac{\langle G^{QQ}(R) \rangle - 2\langle Q \rangle \langle G^{Q1}(R) \rangle + \langle Q \rangle^2 \langle n(R) \rangle}{\langle n(R) \rangle}. \tag{13}$$

We should note that we have made a slight generalization of equation (8) in that the denominator is now the expectation of $n(R)$. An alternative approach would be to bring the $1/n(R)$ inside the expectation value of each term. There is no clear consensus as to which approach is more appropriate, and the difference between the two is small.

IV RESULTS AND CONCLUSIONS

We produced about 2000 independent field configurations for each of several different values of κ_0 for $V = 4000$ simplices and $N = 5$ vector fields. On each of the configurations, we measured connected correlators for the gauge field quantities

$$(A_{ij} + A_{jk} + A_{ki})^2 \tag{14}$$

and

$$(A_{ij})^2. \tag{15}$$

The former is the lattice version of $(F_{\mu\nu})^2$, a gauge invariant object. The latter is proportional to A^2 and is not gauge invariant so we fix the gauge prior to each measurement. In Figure 1 (left) we show the three raw functions $G^{QQ}(R)$, $G^{Q1}(R)$, and $n(R)$, defined in equations (10) - (12). These particular curves correspond to $\kappa_0 = 30$, $N = 5$, $V = 4000$. In Figure 1 (right) we have showed the residual connected correlation function as computed by equation (13). Note this process extracts a signal that is four to five orders of magnitude smaller than the raw curves.

Figure 2 shows the decay of the connected correlators for both $(A_{ij} + A_{jk} + A_{ki})^2$ and $(A_{ij})^2$ in a log-log plot. The full function has been truncated after $R = 6$ because statistical fluctuations cause it to go below zero. Nevertheless the initial decay seems consistent with power law decay with a power of about 2.7 for each. This suggests that a massless propagating mode may survive the quantization of the geometry. A massless mode is necessary, though not sufficient for there to be non-trivial infrared physics. The fact that the exponents may suggest that there is some significant mixing with a smaller dimensioned operator.

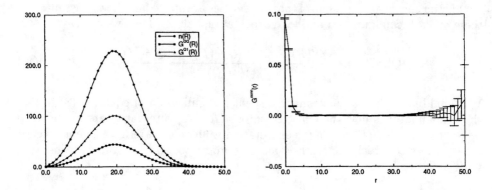

FIGURE 1. The three raw functions needed to compute the connected correlator (left) and the connected correlator as computed by equation (13) (right).

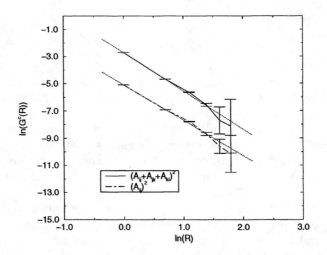

FIGURE 2. A log-log plot of the connected correlators for two gauge field quantities. $(A_{ij} + A_{jk} + A_{ki})^2$ is fit to a line with slope=-2.8 and $(A_{ij})^2$ is fit to a line with slope=-2.6.

REFERENCES

1. Bialas P., and Burda Z., and Krzywicki A., and Petersson B., *Nucl. Phys.* **B 472**, 293–308, (1996).
2. Ambjorn J., and Bialas P., and Jurkiewicz, J. *JHEP* **02**, 005, (1999).
3. Antoniadis I. *Spring Workshop on String Theory*, Trieste, Italy, 1992.
4. Antoniadis I., and Mazur P. O., and Mottola E., *Phys. Lett.* **B 394**,49-56, (1997).
5. Antoniadis I., and Mazur P. O., and Mottola E., *Phys. Lett.* **B 444**,284, (1998).
6. de Bakker B.V., and Smit J., *Nucl. Phys.* **B 454**, 343-356, (1995).

The N-Cosine Model — Algebraic Structures and Integrable Points on the Marginal Manifold

Bogomil Gerganov

Newman Laboratory of Nuclear Studies
Cornell University
Ithaca, NY 14853 — USA

Abstract. We review some algebraic aspects of integrable models in 2 space-time dimensions, such as quantum group and Yangian symmetry. We also discuss the application of perturbed conformal field theory and renormalization group techniques to analyzing the integrability properties of 2 dimensional quantum field theories by probing their underlying algebraic structure. These methods are then used to investigate the integrability of the N-cosine model, a multi-field generalization of the sine-Gordon model, and integrable points are found on a manifold of the parameter space where the interaction becomes marginal.

I INTRODUCTION

The study of integrable systems in 2 space-time dimensions has revealed important new relations between quantum field theory and mathematics. The rich algebraic structures underlying integrability have become research subjects on their own, providing at the same time powerful techniques for exploration of various nonperturbative aspects of QFT.

In Section 2 of our talk we briefly introduce the notion of integrability and outline the method of conformal perturbation theory (CPT) as an important tool for analyzing the properties of 2-dimensional QFT by building conserved quantities. We also show how the formalism of CPT can be applied to studying the renormalization group flows of QFTs with marginal perturbations. A brief discussion of quantum group and Yangian symmetries is also included in this section.

In Section 3 we apply the above techniques to study the integrability properties of the N-cosine model. Models of this type have rather complicated behavior since they depend on many dimensionless coupling parameters and are integrable only on submanifolds of the parameter space. We focus more closely on a submanifold where the N-cosine perturbation becomes marginal and show by RG analysis the presence of integrable points for $N=$ 2, 3, and 4.

II METHODS FOR STUDYING INTEGRABILITY PROPERTIES OF 1+1 DIMENSIONAL QUANTUM FIELD THEORY

Because of the strongly restrictive features of particle kinematics in 2-dimensional space-time, there exist scattering theories which allow precise computation of all elements of the exact particle S-matrix. A QFT in 2 space-time dimensions is integrable if there exists an infinite set of independent, mutually commuting integrals of motion of increasing tensorial rank. The existence of such conserved quantities leads to elasticity and factorizability of the S-matrix. These 2 properties, together with the Yang-Baxter equation and the requirements for analyticity, unitarity, and crossing symmetry, for many 2D QFTs provide enough constraints to compute the S-matrix exactly[1].

A Perturbed Conformal Field Theory (CPT)

A powerful technique for building conserved quantities in 2D models is Zamolodchikov's formalism of perturbed conformal field theory (CPT) [4]. By treating a 2D QFT as a perturbed CFT, one can make use of conformal invariance to find an infinite number of conserved quantities in the conformal theory. It is then possible to study which (if any) of these conservation laws survive the perturbation. Zamolodchikov's paper [4] provides an easy way of computing the conserved current densities explicitly.

In CFT any conserved density T_{s+1} is a holomorphic function and $\partial_{\bar{z}} T_{s+1} = 0$. In the perturbed QFT that is no longer true and we can compute $\partial_{\bar{z}} T_{s+1}$, using Zamolodchikov's formula [4, eq.(3.14)]:

$$\partial_{\bar{z}} T_{s+1} = \lambda \oint_z \frac{d\zeta}{2\pi i} \Phi(\zeta, \bar{z}) T_{s+1}(z) \tag{II.1}$$

If the RHS of (II.1) can be expressed as a total ∂_z-derivative of some local operator Θ_{s-1}, there exists a spin s conservation law surviving in the perturbed QFT:

$$\partial_{\bar{z}} T_{s+1} = \partial_z \Theta_{s-1}, \qquad \partial_z \overline{T}_{s+1} = \partial_{\bar{z}} \overline{\Theta}_{s-1}, \tag{II.2}$$

T_{s+1} and Θ_{s-1} being the quantum conserved densities. Using an insightful counting argument Zamolodchikov showed [4, p.650] that if the perturbing operator $\Phi(z,\bar{z})$ is relevant (i. e. if $[\lambda] < 2$), the perturbation series in λ if finite. In fact for many 2D QFTs the first order in CPT is exact[2].

---

[1]) For a review of the properties of the S-matrix in 2D see Ref. [1]. Classical and quantum integrability in 2-dimensional QFT were also discussed in a talk presented by the author at MRST'98, published in the Proceedings [2].

[2]) For a review of perturbed CFT see Ref. [4]. The application of the method to analyzing the integrability properties of 2D QFT models was also discussed in a talk presented at MRST'98, published in the Proceedings [3].

B Renormalization á la CPT

The formalism of perturbed CFT also provides a convenient way of performing renormalization group (RG) analysis and for calculating beta functions in 2D QFT. Let us consider a conformal field theory perturbed by some marginal operators $\mathcal{O}_i(z,\bar{z})$:

$$S = S_{\text{CFT}} + \frac{1}{4\pi}\int d^2z \sum_i \lambda^i \mathcal{O}_i(z,\bar{z}) \ . \tag{II.3}$$

Zamolodchikov has shown [18] that the beta functions for the couplings λ^i can be computed by examining the singularities in the operator product expansions (OPEs) of the perturbations $\mathcal{O}_i(z,\bar{z})$ and by regularizing them through introducing a cutoff μ. For an OPE of the type

$$\mathcal{O}_j(w,\bar{w})\mathcal{O}_k(z,\bar{z}) \sim C^i_{jk}\frac{\mathcal{O}_i(z,\bar{z})}{|w-z|^2} \ , \tag{II.4}$$

the divergence is logarithmic and the beta functions to second order are

$$\beta_{\lambda^i} \equiv \frac{d\lambda^i}{d\log\mu} = \frac{1}{2}C^i_{jk}\lambda^j\lambda^k + o^3(\lambda^i) \ . \tag{II.5}$$

If the OPEs of the perturbations in (II.3) do not close on the set $\{\mathcal{O}_i(z,\bar{z})\}$, the action (II.3) does not define a consistent QFT. In this case renormalization requires that new operators be added to the Lagrangian until $\{\mathcal{O}_i(z,\bar{z})\}$ becomes closed OPE algebra of the type (II.4). We refer to this procedure by saying that 'new terms are generated under renormalization'.

Let us now consider a type of perturbing operator, which will be particularly important in our subsequent discussion of the N-cosine model. This is the so called 'current-current perturbation',

$$S = S_{\text{CFT}} + \frac{\lambda}{4\pi}\int d^2z \mathcal{O}(z,\bar{z}) \ , \quad \mathcal{O}(z,\bar{z}) = \sum_{a=1}^{\dim(\mathfrak{g})} J_a(z)\overline{J}_a(\bar{z}) \ , \tag{II.6}$$

where $\{J_a(z)\}_{a=1}^{\dim(\mathfrak{g})}$ are field representations of the generators of some Lie algebra $\mathfrak{g}$ in terms of bosonic vertex operators. $J_a(z)$'s satisfy the OPE[3]

$$J_a(w)J_b(z) \sim \frac{k^2\delta_{ab}}{(w-z)^2} + \frac{f^c_{ab}J_c(z)}{w-z} + \text{Reg.} \tag{II.7}$$

and, similarly, for $\overline{J}_a(\bar{z})$. It is then easy to show that

$$\mathcal{O}(w,\bar{w})\mathcal{O}(z,\bar{z}) \sim C_{AD}\frac{\mathcal{O}(z,\bar{z})}{|w-z|^2} \ , \quad \beta_\lambda \equiv \frac{d\lambda}{d\log\mu} = \frac{1}{2}C_{AD}\lambda^2 + o^3(\lambda) \ , \tag{II.8}$$

where C_{AD} is the Dynkin index of the fundamental representation of the algebra $\mathfrak{g}$.

---
[3] This is the general OPE for a current algebra at level k. In this paper, however, k is always 1.

C Quantum Group and Yangian Symmetry

As mentioned earlier, the presence of an infinite series of mutually commuting conserved quantities in 2D QFT guarantees its integrability. It is still a nontrivial task, however, to actually compute the exact S-matrix of a theory. Sophisticated techniques, such as the Bethe Ansatz and the quantum inverse scattering method (QISM), have been developed for this purpose. For an extensive introduction to this subject we refer the reader to [5] and the references therein.

Here we would like to briefly mention another approach to computing exact S-matrices — the method of non-local charges. The study of a variety of integrable models during the last decade has revealed in many of them the presence of non-local conserved quantities satisfying a highly nontrivial algebra — a q-deformed affine Lie algebra, or quantum group symmetry, $U_q(\widehat{\mathfrak{g}})$. One essential feature of the non-local charges is their nontrivial action on multiparticle states (nontrivial comultiplication). Since the S-matrix must commute with the non-local charges, their nontrivial comultiplication imposes very stringent constraints on the S-matrix which imply the Yang-Baxter equation and together with the conditions of analyticity, unitarity, and crossing-symmetry are enough to solve for the S-matrix exactly. In addition, quantum group symmetry provides also clues about the particle spectrum of the theory since a particle multiplet should transform under some irreducible representation of $U_q(\widehat{\mathfrak{g}})$.

In the subsequent discussion we will encounter another type of infinite dimensional extension of a Lie algebra — a Yangian $Y(\mathfrak{g})$ over a Lie algebra $\mathfrak{g}$. Bernard and LeClair showed [19] that in the case of sine-Gordon model and affine Toda theories the algebra of the non-local charges becomes Yangian in the marginal limit. The non-local charges satisfying $Y(\mathfrak{g})$ also have nontrivial comultiplication which allow finding the exact S-matrix of the theory.

We refer the reader to [19] and the references therein for a discussion of the method of non-local charges and its application to computing exact S-matrices.

III THE N-COSINE MODEL

The N-cosine model is a N-field generalization of the sine-Gordon model (SG). Multi-field generalizations of the SG model have been previously studied in a number of works [8–17]. The Euclidean action of the model is

$$S = \frac{1}{4\pi} \int d\tau dx \left[\frac{1}{2} \sum_{i=1}^{N} (\partial_\mu \Phi_i)^2 + \lambda \prod_{i=1}^{N} \cos(\beta_i \Phi_i) \right] . \qquad \text{(III.1)}$$

A Integral of Motion of Lorentz Spin 3

A calculation to first order in conformal perturbation theory [4] showed [6] that the model (III.1) possesses a conserved current of Lorentz spin 3, when the couplings lie on the submanifold of the parameter space determined by the constraint

$$\beta_1^2 + \beta_2^2 + \ldots + \beta_N^2 = 2 \ . \tag{III.2}$$

The most general Ansatz for T_4 in the N-field case, including all possible local fields of dimension 4 which respect the symmetries of the action (III.1), is [4]

$$T_4 = \sum_{i=1}^{N} \left[a_i (\partial_z^2 \phi_i)^2 + b_i (\partial_z \phi_i)^4 \right] + \sum_{i<j} c_{ij} (\partial_z \phi_i)^2 (\partial_z \phi_j)^2 \ . \tag{III.3}$$

Using (II.1) with $s = 3$ to compute $\partial_{\bar{z}} T_4$, we showed that for $N \geq 3$ the RHS of (II.1) can be expressed as a total ∂_z-derivative of some local operator, $\partial_z \Theta_2$, if and only if the coupling parameters β_i satisfy (III.2).

It is interesting to compare the result for $N \geq 3$ with the ($N=2$)-case. As shown in [14] and [16], a quantum conserved current (to first order in CPT) exists on 3 distinct submanifolds in (β_1, β_2)-parameter space:

$$\beta_1^2 - \beta_2^2 = 0 \ , \tag{III.4}$$

$$\beta_1^2 + \beta_2^2 = 1 \ , \tag{III.5}$$

$$\beta_1^2 + \beta_2^2 = 2 \ . \tag{III.6}$$

The first manifold (III.4) is trivial: when $\beta_1^2 = \beta_2^2$, the double cosine model decouples into 2 sine-Gordon models and, of course, is integrable both classically and quantum mechanically. On the second manifold (III.5) the exact S-matrix has been found by Lesage et al. [15], using the method of non-local charges. The integrability of the model on the third manifold (III.6), to the best of our knowledge, has not been thoroughly investigated.

The fact that T_4 is conserved on the manifold (III.2) for arbitrary N is encouraging. One could hope that it is due to some yet undiscovered symmetry of the theory (III.1). The result is, however, challenged by the following dimensional argument: the quantity β^2 (sum of the couplings squared) is exactly the scaling dimension of the perturbing operator in (III.1) and when $\beta^2 = 2$, this operator becomes marginal ($[\lambda] = 0$). As a result Zamolodchikov's counting argument [4, p.650] is weak in this case and first order CPT can no longer be claimed to give exact results. Therefore we need to use other methods to investigate further the integrability of the model on the marginal manifold. As we will show below, renormalization group analysis, conducted using the technique outlined in Section II.B, allows us to related the model (III.1) to known integrable models at special points on the marginal manifold.

---

[4] See also [9, 10, 15, 16].

B Renormalization on the Marginal Manifold and Integrable Points for $N = 2$

Following Zamolodchikov's prescription (Cf. Section II.B), we performed a renormalization group analysis of the model (III.1) with $N=2$ on the marginal submanifold $\beta_1^2 + \beta_2^2 = 2$. There are 5 distinct regimes, depending on the value of the quantity $|\beta_1^2 - \beta_2^2|$:

$$
\begin{aligned}
&1. \quad |\beta_1^2 - \beta_2^2| = 0 \ ; \\
&2. \quad 0 < |\beta_1^2 - \beta_2^2| < 1 \ ; \\
&3. \quad |\beta_1^2 - \beta_2^2| = 1 \ ; \\
&4. \quad 1 < |\beta_1^2 - \beta_2^2| < 2 \ ; \\
&5. \quad |\beta_1^2 - \beta_2^2| = 2 \ .
\end{aligned}
\qquad \text{(III.7)}
$$

In Case 1 the couplings β_1 and β_2 are equal and the model (III.1) decouples into two sine-Gordon models. In Case 2 the double-cosine model is non-trivial but no new terms are brought in by renormalization. In Case 3 renormalization requires that new marginal terms be brought into the action. In Case 4 the OPE (II.4) has power-law divergences which lead to a generation of a mass term under renormalization. In Case 5 one of the couplings vanishes while the other has a value of $\pm\sqrt{2}$ and (III.1) is reduced to a single SG model at its marginal point and a free boson field. (The marginal limit of the SG model is discussed in details in [19].)

Let us now turn to the cases 1, 3, and 5 of (III.7) in which the quantity $|\beta_1^2 - \beta_2^2|$ has integer values. The constraints (III.7) in these case specify points on the marginal submanifold (III.2) where the double-cosine interaction, together with the additional terms brought in by renormalization, can be written as a current-current perturbation to the level 1 $\mathfrak{g}$ current algebra[5]:

$$
\frac{\tilde{\lambda}}{4\pi}\int d^2z \left\{ \sum_{\alpha_j > 0} \left[E^{\alpha_j}(z)\overline{E}^{-\alpha_j}(\bar{z}) + E^{-\alpha_j}(z)\overline{E}^{\alpha_j}(\bar{z}) \right] + \sum_{i=1}^{\mathrm{rank}(\mathfrak{g})} H^i(z)\overline{H}^i(\bar{z}) \right\}
\qquad \text{(III.8)}
$$

with $\tilde{\lambda} \equiv \frac{\lambda}{4}$. $\{\alpha_j\}$ are the roots and $E^{\alpha_j}(z)$ are the vertex representations of the corresponding generators of some Lie algebra $\mathfrak{g}$ in the Cartan-Weyl basis[6]. As we showed in [6], even though the double-cosine perturbation (III.1) contains only the generators corresponding to some of the roots of $\mathfrak{g}$, the generators corresponding to remaining roots as well as the Cartan generators of $\mathfrak{g}$ are generated under renormalization, thus leading to the model (III.8). At level 1 the model (III.8) is simply the $\mathfrak{g}$-invariant Gross-Neveu (GN) model which is known to be integrable and its S-matrix has been computed for most classical Lie algebras [20, 21]. Studying the

---

[5] Also called the $\mathfrak{g}$ Wess-Zumino-Witten (WZW) model at level 1.
[6] For a detailed discussion of the representations of Lie algebras in terms of bosonic vertex operators see for instance [22, 23].

marginal limit of the affine Toda theories, [19] showed that models of GN type possess Yangian symmetry, which provides an independent way of computing the S-matrix.

The symmetry algebra $\mathfrak{g}$ turns out to be $\mathfrak{su}(2) \oplus \mathfrak{su}(2)$ in Case 1 (2 decoupled SG models at their marginal points) and $\mathfrak{su}(2)$ in Case 5 (a single marginal SG model and a decoupled free scalar field). The constraint of Case 3 specifies new integrable points on the marginal manifold that have not been identified before. The symmetry algebra in this case is $\mathfrak{g}=\mathfrak{su}(3)$. The beta function of $\tilde{\lambda}$ in the Cases 1, 3, and 5 is given by (II.8) with $C_{AD}^{\mathfrak{su}(2)} = 4$ and $C_{AD}^{\mathfrak{su}(3)} = 6$.

Case 2 and Case 4 correspond to the segments of the marginal circle between the $\mathfrak{g}$-invariant points. In these cases the model cannot be written as a current-current perturbation.

Case 2 corresponds to the segments interpolating between the $\mathfrak{su}(2) \oplus \mathfrak{su}(2)$-invariant and the $\mathfrak{su}(3)$-invariant points. Although the model in this regime is non-trivial and its properties could be further investigated, the following simple argument makes integrability look very unlikely: at the $\mathfrak{su}(2) \oplus \mathfrak{su}(2)$-invariant points the particle spectrum of the model consists of 4 particles: 1 soliton and 1 anti-soliton for each SG model. On the other hand, the particle spectrum of the $\mathfrak{su}(3)$-invariant GN model consists of 6 particles — belonging to the two fundamental representations, $\mathbf{3}$ and $\bar{\mathbf{3}}$, of $\mathfrak{su}(3)$. Therefore it is not conceivable, just by counting the degrees of freedom, that the particle spectrum could be smoothly deformed from the first type of points to the second.

Case 4 corresponds to the segments on the marginal circle, interpolating between the $\mathfrak{su}(3)$-invariant and the $\mathfrak{su}(2)$-invariant points. As mentioned before, in this case renormalization leads to a generation of a massive term in the action. This new massive perturbation breaks the conservation of the spin 3 current (II.2) even to first order in CPT. Therefore, in this regime the model is not integrable.

C Generalization to Arbitrary N

The renormalization group analysis of Section III.B can be generalized to arbitrary N. As in the 2-field case, the marginal N-cosine model can be written as a current-current perturbation to some current algebra $\mathfrak{g}$ at special points on the marginal N-sphere (III.2), where the values of the coupling parameters β_i are consistent with the root structure of $\mathfrak{g}$. We found [6] that such marginal points exist for $N=3$ and $N=4$. For $N>4$ the marginal N-cosine model cannot be written as a current-current perturbation.

In the $N=3$ case the 3-cosine action is equivalent to a current-current perturbation of the $\mathfrak{su}(4)$ current algebra at the points for which the values of the couplings β_i are specified by the equations:

$$(\beta_1^2, \beta_2^2, \beta_3^2) = \left(\frac{1}{2}, 1, \frac{1}{2}\right), \quad \left(1, \frac{1}{2}, \frac{1}{2}\right), \quad \left(\frac{1}{2}, \frac{1}{2}, 1\right). \tag{III.9}$$

For $N=4$ the $\mathfrak{g}$-invariant points on the marginal manifold must have coordinates consistent with

$$(\beta_1^2, \beta_2^2, \beta_3^2, \beta_4^2) = \left(\frac{1}{2}, \frac{1}{2}, \frac{1}{2}, \frac{1}{2}\right) . \quad (III.10)$$

At the above points the 4-cosine action is equivalent under renormalization to a current-current perturbation of the level 1 $\mathfrak{so}(8)$ current algebra.

In both of the above cases the beta function of $\widetilde{\lambda}$ is given by (II.8) with $C_{AD}^{\mathfrak{su}(4)} = 8$ and $C_{AD}^{\mathfrak{so}(8)} = 12$.

Finally, it is interesting to note that there are two other classes of models, the deformed Gross-Neveu models and the affine Toda theories, which have the same marginal limiting behavior as the N-cosine model (III.1) at its $\mathfrak{g}$-symmetric marginal points[7].

IV CONCLUSIONS

We discussed the integrability properties of the N-cosine model as an illustration of various techniques for studying 2-dimensional QFT models. We established that, for arbitrary N, the model possesses a current of Lorentz spin 3, conserved to first order in conformal perturbation theory on a submanifold of the parameter space where the interaction becomes marginal. For $N=2$, 3, and 4, we found points on the marginal manifold at which the model can be written as a current-current perturbation to some current algebra $\mathfrak{g}$. Such marginal models are of the Gross-Neveu type and are therefore integrable. In the 2-field case we further argued that the $\mathfrak{g}$-invariant points exhaust all integrable cases on the marginal manifold.

ACKNOWLEDGMENTS

I would like to thank André LeClair for support and advice and for numerous useful discussions and insights. I am also thankful to Marco Ameduri and Zorawar Bassi for many discussions and useful ideas and to the organizers of MRST'99 for their hospitality and for creating an excellent atmosphere during the conference.

REFERENCES

1. Zamolodchikov, A. B., in *Soviet Scientific Reviews*, Sec. **A**, ed. Khalatnikov, I. M., New York: Harwood Academic Publishers, 1980, vol. 2, pp. 1–40.
2. Gerganov, B., "Integrability in Classical and Quantum Field Theory and the Bukhvostov–Lipatov Model," in *AIP Conference Proceedings* **452**, *Toward the Theory of Everything: MRST'98*, eds. Cline, J. M., et al., Woodbury, New York: American Institute of Physics, 1998, pp. 144–152.

---

[7] See Ref. [6] for a more extensive discussion.

3. Ameduri, M., "Perturbed Conformal Field Theory: A Tool for Investigating Integrable Models," in *AIP Conference Proceedings* **452**, *Toward the Theory of Everything: MRST'98*, eds. Cline, J. M., et al., Woodbury, New York: American Institute of Physics, 1998, pp. 195–200.
4. Zamolodchikov, A. B., *Adv. Studies in Pure Math.* **19**, 641 (1989).
5. Korepin, V. E., and Bogoliubov, N. M., and Izergin, A. G., *Quantum Inverse Scattering Method and Correlation Functions*, Cambridge: Cambridge University Press, 1993.
6. Gerganov, B., "Integrable Marginal Points in the N-Cosine Model," hep-th/9905214.
7. Shankar, R., *Phys. Lett.* **B102**, 257 (1981).
8. Bukhvostov, A. P., and Lipatov, L. N., *Nucl. Phys.* **B180**, 116 (1981).
9. Fateev, V. A., *Phys. Lett.* **B357**, 397 (1995).
10. Fateev, V. A., *Nucl. Phys.* **B473**, 509 (1996).
11. Baseilhac, P., Grangé, P., and Rausch de Traubenberg, M., *Mod. Phys. Lett.* **A13**, 2531 (1998).
12. Saleur, H., and Simonetti, P., *Nucl. Phys.* **B535**, 596 (1998).
13. Ameduri, M., and Efthimiou, C. J., *J. Nonl. Math. Phys.* **5**, 132 (1998).
14. Lesage, F., Saleur, H., and Simonetti, P., *Phys. Rev.* **B56**, 7598 (1997).
15. Lesage, F., Saleur, H., and Simonetti, P., *Phys. Rev.* **B57**, 4694 (1998).
16. Ameduri, M., Efthimiou, C. J., and Gerganov, B., "On the Integrability of the Bukhvostov–Lipatov Model," hep-th/9810184.
17. Saleur, H., "The Long Delayed Solution of the Bukhvostov–Lipatov Model," hep-th/9811023.
18. Zamolodchikov, A. B., *Sov. J. Nucl. Phys.* **46**, 1090 (1987).
19. Bernard, D., and LeClair, A., *Commun. Math. Phys.* **142**, 99 (1991).
20. Karowski, M., and Thun, H. J., *Nucl. Phys.* **B190**, 61 (1981).
21. Ogievetsky, E., and Reshetikhin, N. Yu., Wiegmann, P., *Nucl. Phys.* **B280** [FS18], 45 (1987).
22. Di Francesco, P., Mathieu, P., and Sénéchal, D., *Conformal Field Theory*, New York: Springer, 1996.
23. Fuchs, J., *Affine Lie Algebras and Quantum Groups*, Cambridge: Cambridge University Press, 1992.

Coherent States in High-Energy Physics

C.S. Lam

Department of Physics, McGill University
3600 University St., Montreal, QC, Canada H3A 2T8

Abstract. The amplitude for emitting n bosons factorizes into the product of n single-boson emission amplitudes, if the source is energetic and abelian. If it is energetic but *non-abelian*, the amplitude is given by a sum of factorized *quasi-particle* amplitudes. A quasi-particle is made up of an arbitrary number of bosons, but couples to the source like a single one. Factorization is related to coherence, and it allows computation of subleading contributions not obtainable by usual means. Its importance is illustrated in two applications: to solve the baryon problem in large-N_c QCD, and to obtain a total cross section satisfying the Froissart bound.

I INTRODUCTION

We found a quasi-particle state of gluons whose existence has eluded detection all these years. In this talk I will discuss how that comes about, and what use we can make of it. A quasi-particle is made up of an arbitrary number of gluons, but it couples to their high-energy source like a single one: as a colour-octet object whose emission preserves helicity of the source. Quasi-particles emerge naturally as a result of factorization and coherence. They are present in all non-abelian theories, including the Yukawa theory of nucleons and pions, and not just QCD.

By a high-energy source, I mean a source with large *total* energy. The source may be a highly relativistic particle with a small mass, or a non-relativistic particle with a very large mass. For simplicity, I shall refer to these two cases respectively as a relativistic source and a non-relativistic source.

By a non-abelian theory, I mean one in which the spin and/or the internal quantum numbers of the high-energy source can be altered by the emission of bosons. In the case of pions emitted from a massive non-relativistic nucleon, it is the spin and the isospin of the nucleon that are affected by the emission. In the case of QCD it is the colour of the source. But in the case of photons emitted from a relativistic electron, neither the charge nor the helicity of the electron is changed, so in this case the source is abelian.

After sketching the origin of the quasi-particle, and its connection with factorization and cohernece, I will discuss two cases in which it makes its presence felt. I believe these applications barely scratch the surface, and the importance of these quasi-particles goes far beyond these two examples, but exactly in what way remains to be seen.

II THE EMERGENCE OF QUASI-PARTICLES

Consider the tree diagram Fig. 1(a), in which n bosons are emitted from an energetic source via vertex factors V_i. The source is assumed to be energetic so that recoils suffered from the emissions can be ignored. As a result, its transverse position $x_\perp$ is fixed, and so is the $x_\perp$ of every boson emerging from it. In the case of a relativistic source, we may assume it to move parallel to the z-axis near the speed of light, hence its $x^- \equiv x^0 - x^3$ coordinate is fixed, thereby determining also the x^- coordinate of all the emerging bosons. For a non-relativistic source, its x^3 coordinate and that of the bosons are fixed. All in all, three out of four coordinates of all the off-shell bosons are identical. If the fourth one is also the same for all the bosons, a coherent state will emerge. It turns out that non-trivial inputs are required to determine this fourth coordinate to achieve coherence.

It is necessary to invoke Bose-Einstein symmetry, which in this case simply means summing up the $n!$ permuted tree amplitudes. For abelian sources, the vertices V_i's can be regarded simply as numbers. In that case we will show that every boson of

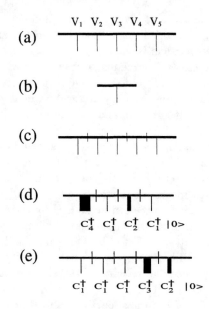

FIGURE 1. (a). A Feynman tree amplitude with vertex factors V_i; (b). A one-boson amplitudes with on-shell sources; (c). The factorization of a Bose-Einstein symmetrized amplitude into the product of one-boson amplitudes; (d) and (e). Two separate factorization into quasi-particle amplitudes.

the BE-symmetrized amplitude has its momentum component $k^- = k^0 - k^3 = 0$ if the source is relativistic, and $k^0 = 0$ if the source is non-relativistic. This is the fourth coordinate we are after for a coherent state.

The conclusion follows as a result of factorization. After the summation, each boson is allowed to be emitted anywhere along the tree, irrespective of the location of the others. Hence the n-boson amplitude is factorized into a product of n single-boson emission amplitudes. This is depicted in Fig. 1(c), where a vertical cut on the tree indicates factorization. For a relativistic source that is on-shell, its momentum component $p^- = p^0 - p^3 = 0$, so by momentum conservation (see Fig. 1(b)) $k^- = 0$ for every boson as claimed. For a non-relativistic source that is on-shell, p^0 is fixed at its on-shell mass M, so by momentum conservation $k^0 = 0$. Note that neither conclusion is valid for the boson momenta in a Feynman diagram like Fig. 1(a), where off-shell sources are involved. In that case, uncertainty in energy prevents p^- or p^0 to be fixed, so it is not true that k^- or k^0 is zero. Bose-Einstein symmetrization and the resulting factorization are crucial to reach these conclusions.

Note also that this kind of coherent state is very different from those encountered at low temperatures, where Bose-Einstein condensation may occur. The coherent state we have is described by a mixture of spatial and momentum coordinates, and it is not an energy eigenstate. What is 'cold' in the present context is the lack of recoil, instead of the lack of thermal fluctuation. Hence the physics outcome between the two are completely different as well.

For non-abelian sources this simple factorization is no longer valid. The vertex factors V_i fail to commute with one another, so correction terms involving their commutators must be added [1]. It turns out that each of these correction terms is still factorizable, but generally into products of *quasi-particle* amplitudes instead of single-boson amplitudes. In other words, it is the k^- or the k^0 coordinates of the quasi-particles as a whole that are zero, but not the individual bosons within each quasi-particle. As remarked before, a quasi-particle may consist of any number k of bosons, but instead of coupling to the source via the product of vertex factors $V_1 V_2 \cdots V_k$, it does so via the nested commutator $[V_1, [V_2, [\cdots, [V_{k-1}, V_k] \cdots]]]$. In the case of QCD when V_i are colour matrices, the nested commutator is given by a linear sum of colour matrices, so the quasi-particle couples just like a colour-octet object. Moreover, since each gluon making up the quasi-particle does not flip the helicity of the source, neither will the quasi-particle.

Exactly how each correction term factorizes depends on the permutation. I list here three examples for $n = 8$: [1|2|3|4|5|6|7|8], [8521|3|74|6], [1|2|3|864|75], in which a vertical bar indicates the position where factorization takes place. The general rule is simply that a bar should be put behind a number iff no number to its right is smaller than it. The first example is identical to abelian factorization. In the second example, the source emits a quasi-particle of four gluons, one of two gluons, and two with one gluon each (a quasi-particle with one gluon is just a gluon). In the third example, there are three quasi-particles with one gluon each, one with three gluons, and one with two gluons. The last two examples involve nested commutators so they will not be present for abelian sources. In fact, other than the first example, no other permutation can contribute in the case of an abelian source for exactly the same reason.

Letting Q denote a quasi-particle, the general structure of each factorized amplitudes is therefore of the form

$$[Q|Q|\cdots|Q], \qquad (1)$$

where the different quasi-particles Q appearing in this equation may consist of different number of gluons.

III COMPOSITE SOURCE

Suppose the source is made up of N constituents, each capable of emitting a boson via the vertex $V_i = \psi^\dagger t_i \psi$, where ψ and $\psi^\dagger$ are the annihilation and creation operators for the constituents. For example, the source may be a nucleus with N nucleons, or a nucleon with N quarks. Being a one-body operator, the matrix elements of V_i is expected to be of order N. Being a k-body operator, the matrix element of a product of k V_i's is expected to be of order N^k. In contrast, the nested commutator of k V_i's is of the form $\psi^\dagger T \psi$, with T given by the nested commutator of the t_i's, so it is still a one-body operator and its matrix element is proportional to N. If the an n-boson amplitude in (1) contains p quasi-particles, then that term is of order N^p, with $p \leq N$, and not N^n that each Feynman tree diagram is expected to have. Except for the identical permutation whose amplitude factorizes completely as in $[1|2|3|\cdots|n]$, so that $p = n$, all the others have $p < n$ and hence contribute subdominantly when $N \gg 1$. The smaller p is the less it contributes. If for some reason all the terms with $n \geq p \geq p_0 + 1$ vanish, then the amplitude is of order N^{p_0}. It can still be computed easily from the quasi-particle amplitudes with $p \leq p_0$, but it is extremely difficult to calculate it directly from Feynman tree diagrams, especially when $p^0 \ll n$. To do so we must compute each Feynman diagram to the subleading order N^{p_0} before a finite sum can be obtained upon summation, a highly non-trivial task.

Such a behaviour indeed happens in the process $\pi + \mathcal{N} \to (n-1)\pi + \mathcal{N}$, calculated in large-$N$ QCD [2]. In that case, the nucleon $\mathcal{N}$ consists of N quarks. Its mass is of order N so it is a non-relativistic energetic source. The effective Yukawa interaction $t = g(\vec{\sigma} \cdot \vec{k})(\vec{\tau} \cdot \vec{\pi}(\vec{k}))$ is non-abelian because it flips the spin and the isospin of the nucleon. Each Feynman tree amplitude is of order $N^n/\sqrt{N}^n$ because it consists of n vertices and the propagators are of order 1. A normalization factor $1/\sqrt{N}$ per pion is put in as usual [2]. The amplitude is huge for every $n > 0$ in the limit $N \gg 1$. In this strong-coupling limit one might think that very little could be said about the reaction, and certainly the Feynman-diagram description is useless even when loops are included. Yet the phenomenology of baryons in large-N QCD is very successful in describing nature [3]. What happens is that when the $n!$ permuted diagrams are summed up, a tremendous amount of cancellation takes place, so that the final n-pion amplitude is of order $N^{1-n/2}$ rather than $N^{n/2}$ of the individual diagrams. The total pion-nucleon amplitudes now become weak for $n > 2$, so we can understand why loops are not needed and why phenomenology can be successful. In order to prove this cancellation in a brute-force way, each diagram must be calculated down to $(n-1)$ subleading orders, for at the end everything else above it will be cancelled in the sum. This is quite an impossible task for large n, and this is where the advantage of the factorized formula (1) shows up [4]. If the number p of quasi-particle amplitudes in (1) is larger than 1, then it vanishes because of energy conservation. In that case one of these p factorized components must consist of only outgoing pions, which violates energy conservation since the nucleons are on-shell. As a result we are left with only terms with $p = 1$, whose matrix element is $N/\sqrt{N}^n = N^{1-n/2}$, as claimed.

IV DAMPING EXPLOSIVE TOTAL CROSS SECTIONS

Total cross section can be obtained from the forward elastic amplitude via the optical theorem. This amplitude is difficult to compute even assuming the coupling constant α_s to be small, for at high cm energy complicated loop diagrams must be included. This is so because each time we add a loop to the diagram, we add an extra factor of α_s but the loop integration may also produce an extra $\ln s$. Thus a diagram of order $2n+2$ may give a contribution proportional to $\alpha_s(\alpha_s \ln s)^n$, which is of order α_s if $\xi \equiv \alpha_s \ln s = O(1)$. This is why diagrams of all orders must be included.

Computing multi-loop diagram is a difficult task which can be accomplished only with suitable approximations. In the *leading-log approximation*, which keeps only the lowest power of α_s while keeping the variable ξ fixed, the total cross section so computed is proportional to $\alpha_s^2 \exp(4\alpha_s \ln s \ln 2N_c/\pi)$. This is the famous BFKL formula [5]. With $\alpha_s \simeq 0.19$, for example, the cross section according to this formula grows with energy like $s^{0.5}$. At this rate the size of a proton becomes ten times the size of the Uranium nucleus at LHC energy, and one hundred times its size at 10^{20} eV, the highest energy cosmic-ray reaching earth. For better or for worse, this alarming growth is not realized. In fact, Froissart bound forbids the total cross section to grow faster than $\ln^2 s$ at asymptotic energies. The theoretical challenge then is how to produce sufficient amount of corrections in QCD to satisfy the Froissart bound. Since the BFKL computation already includes all the important contributions in the leading-log approximations, *viz.*, all terms of order α_s^2 when ξ is kept fixed, clearly subleading terms of order α^m with $m \geq 3$ are needed for the Froissart bound. As explained in the last section, the factorization formula (1) is capable of extracting subleading terms N^p for $p < n$ in that case. Similarly, (1) can be used to extract subleading terms α_s^m with $m \geq 3$ [6]. The result is shown in Fig. 1(b), where the thick vertical lines represent quasi-particles, and the thin vertical cuts on the two horizontal lines represent factorization. It can be shown that an amplitude with p vertical quasi-particle lines is of order α_s^p; this is analogous to the situation of the last section in which an amplitude with p quasi-partilces is proportional to N^p.

For $\alpha_s \ll 1$ and $\xi = O(1)$, the dominant scattering amplitude comes from diagrams with $p = 1$ (alone), indicated in Fig. 2(b) by δ_1. Remebering that each quasi-particle carries an octet colour, we conclude that the dominant amplitude is a colour-octet amplitude (in the t-channel), whose magnitide is $O(\alpha_s)$. It has been known long ago [7] that the dominant amplitude is obtained by the exchange of a colour-octet Reggeon, so from these two equivalent descriptions we can identify the quasi-particle with the Reggeon. What has thus been achieved here is an algebraic characterization of the Reggeon, as the colour-octet object obtained through factorization and coherence, rather than a pole in the angular momentum plane as is usually defined.

A quasi-particle in QCD is not the same as a gluon, but a quasi-particle in QED is identical to a photon because all the nested commutators vanish. This distinction is ultimately the reason why gluons reggeize but photons do not.

Total cross section is related to the forward part of the *elastic* scattering amplitude, so only the exchange of colour-singlet object contributes to it. The dominant amplitude then comes from the exchange of two interacting Reggeons, indicated by δ_2 in Fig. 2(b), or two non-interacting Reggeons δ_1^2. The result is of order α_s^2, and it is the BFKL Pomeron [5], which as mentioned before violates unitarity. s-channel

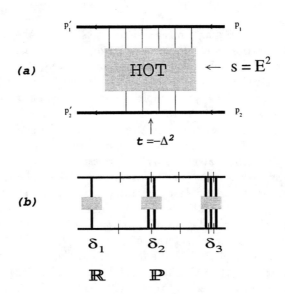

FIGURE 2. (a). A two-particle collision diagram showing a hot central region and a 'cold' peripheral region; (b). Factorization of the scattering amplitude into quasi-particle amplitudes. An amplitude with p quasi-particle exchanges can be shown to be of order α_s^p if $\alpha)s \ln s$ is of order 1.

unitarity and the Froissart bound are restored when we incorporate the singlet part of p-Reggeon exchanges, with all $p \geq 3$ included. For virtual-photon proton total cross section, as measured at HERA [8], this results in a shallower growth of total cross section with energy for a smaller virtuality Q^2 of the virtual photon, as shown in Fig. 3.

For details see Ref. [6].

Finally one might ask why coherence should have anything to do with high-energy collisions. After all, the centre of collision will be hotter than the centre of a star (Fig. 2(a)), whereas Bose-Einstein coherence usually happens at low temperature. The answer is that although the centre is hot, the peripheral regions are 'cold', and that is sufficient to produce factorization as indicated in Fig. 2(b).

REFERENCES

1. Lam C.S., and Liu K.F., *Nucl. Phys.* **B483**, 514 (1997); Feng Y.J., Hamidi-Ravari O., and Lam C.S., *Phys. Rev. D* **54**, 3114 (1996); Lam C.S., *J. Math. Phys.* **39**, 5543 (1998).
2. 't Hooft G., *Nucl. Phys.* **B72**, 461 (1974); Coleman S., Erice Lectures (1979); Witten E., *Nucl. Phys.* **B160**, 57 (1979).
3. Dashen R.F., Jenkins E., and Manohar A.V., *Phys. Rev. D* **49**, 4713; *Phys. Rev. D* **51**, 3657 (1995); Dashen R., and Manohar A.V., *Phys. Lett. B* **315**, 425, 438 (1993); Jenkins E., *Phys. Lett. B* **315**, 431, 441, 447 (1993); Luty M.A., and March-Russell J., *Nucl. Phys.* **B426**, 71 (1994); Luty M.A., *Phys. Rev. D* **51**, 2322 (1995).
4. Lam C.S., and Liu K.F., *Phys. Rev. Lett.* **79**, 597 (1997).
5. Lipatov L.N., *Sov. J. Nucl. Phys.* **23**, 338 (1976); Balitskii Ya., and Lipatov L.N., *Sov. J. Nucl. Phys.* **28**, 822 (1978); Kuraev E.A,, Lipatov L.N., and Fadin V.S., Z*Sov. Phys. JETP* **44**, 443 (1976); *ibid.* **45**, 199 (1977).
6. Dib R., Khoury J., and Lam C.S., hep-ph/9902429, to appear in Phys. Rev. D.
7. For a review, see Cheng H., and Wu T.T., *'Expanding Protons: Scattering at High Energies'*, (M.I.T. Press, 1987); Del Duca V., hep-ph/9503226.
8. ZEUS Collaboration, hep-ex/9707025.

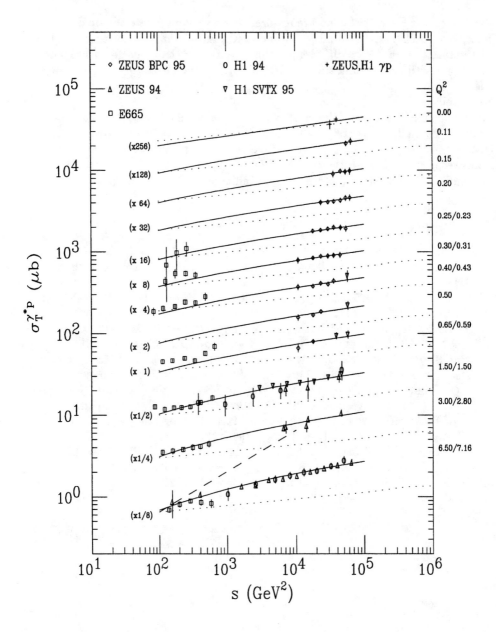

FIGURE 3. Energy dependence of $\gamma^* p$ total cross section as a function of photon virtuality Q^2. Data is from Ref. [8]. The dotted line represents a dependence of $s^{0.08}$ obeyed by all hadronic total cross sections. The dash line gives a $s^{0.5}$ variation predicted by the leading-log BFKL Pomeron, and the solid line is the prediction of the unitary theory in Ref. [6].

The Ising Model on a Fluctuating Disk

Scott V. McGuire, Simon M. Catterall

Department of Physics
Syracuse University
Syracuse, NY 13244

Abstract. We report on simulations of Ising models on dynamically triangulated lattices with disk topology. We discuss the background to these calculations, including the expected scaling behavior with boundary magnetic field and show results confirming the validity of our algorithms.

General Relativity and Quantum Field Theory are two of the most interesting branches of modern physics, so it is natural to wish for a single theory which encompasses the principles of both. In addition to satisfying our aesthetic sense, such a theory may prove useful. For instance, it would resolve the difficulty of coupling quantum matter to classical gravity through Einstein's equation

$$G_{\mu\nu} \equiv R_{\mu\nu} - \frac{1}{2}R g_{\mu\nu} = 8\pi T_{\mu\nu}.$$

The difficulty with this equation is that the gravitational quantities are numbers while the stress-energy tensor with quantized matter is an operator.

Given a theory derivable from an action $S[\psi]$ one can attempt to quantize it using path-integrals. The propagator $\langle \psi_1(x), t_1 | \psi_2(x), t_2 \rangle$ is obtained by summing over all paths with $\psi_1(x, t_1)$ and $\psi_2(x, t_2)$ as boundary conditions.

$$\langle \psi_1(x), t_1 | \psi_2(x), t_2 \rangle = \int e^{iS[\psi]} d[\psi]$$

The propagators can be obtained from the vacuum-vacuum amplitude in the presence of an external source – the partition function:

$$Z = \int e^{iS[\psi]} d[\psi].$$

This oscillatory integral is not well defined so it is replaced by an integral in Euclidean space-time

$$Z = \int e^{-S_E[\psi]} d[\psi].$$

If the solutions to the Euclidean theory are suitably behaved, they can be continued to Minkowski space-time where the physical theory lives. Applying this to GR, one starts with the Euclidean Einstein-Hilbert action

$$S_E[g] = \int \sqrt{g}(-R + 2\Lambda)d^d x$$

and writes

$$Z = \int e^{-S_E[g]} d[g].$$

This attempt runs into several difficulties. First, many different metrics, related by diffeomorphisms, describe the same space-time. Only one metric from each space-time should be included in the integral and it is not clear how to construct a measure for which this is the case. Second, $S_E[g]$ is unbounded from below, so the functional integrals are not damped and are not well defined. This is true even in spaces of Euclidean signature. Third, it is not clear that one can obtain solutions to Minkowski signature problems from Euclidean signature solutions. Fourth, dimensional arguments imply that the theory can not be renormalized. Amid these problems there is a pleasant exceptional case: The prescription works and can be carried out analytically in the case of 2d quantum gravity coupled to conformally invariant matter.

In the dynamical triangulations (DT) approach, one discretizes the integral over all manifolds by replacing it with a sum over triangulated manifolds. These are piecewise linear manifolds produced by gluing together flat d-dimensional simplices with fixed edge lengths – in 2 dimensions for instance, a surface built by gluing together equilateral triangles. This provides a measure in the space of manifolds for the functional integral, ensures that $S_E[g]$ is bounded from below, and keeps the integral finite. In 2 dimensions, on the sphere with conformal matter, this has been shown to agree with analytical results.

We wish to extend previous simulations to a disk topology. This allows us to investigate theories with non-trivial boundary conditions, and corresponds to the situation in Hartle-Hawking "wavefunction of the universe" calculations. Specifically, we have simulated pure gravity on the disk and the Ising model on the disk with and without a boundary magnetic field.

Given a triangulation of the sphere with a given volume (number of triangles), any other triangulation of the same volume can be reached via a series of link flips. If in addition to link flips, we allow insertion of new nodes into triangles and the inverse move – deleting nodes surrounded by only three triangles, we have a set of moves which are ergodic in the space of triangulated spheres. Any disk can be turned into a sphere by attaching all the boundary nodes to a new node. This is a one-to-one mapping between disks and spheres with a marked node. Thus, we can simulate disks by simulating marked spheres. To do this, the above rules are modified by keeping track of a special node on the sphere and not allowing it to be deleted. The following variables describe the geometry:

- n_0^{sphere} The number of nodes in the sphere

- n_1^{sphere} The number of links on the sphere
- n_2^{sphere} The number of triangles on the sphere
- q The number of nodes the special node is connected to, also the boundary length of the disk
- $n_0^{disk} = n_0^{sphere} - 1$ The number of nodes on the disk
- $n_1^{disk} = n_1^{sphere} - q$ The number of links on the disk
- $n_2^{disk} = n_2^{sphere} - q$ The number of triangles on the disk

As the Einstein-Hilbert action is a topological invariant in two dimensions, in principle there is no need to consider it. However, as the number of triangulations of the sphere grows exponentially with the volume, a naive simulation of disks without a constraint on volume would lead to disks of infinite volume. To remedy this, we add to the action a cosmological constant term $\exp\{\kappa_2 n_2^{sphere}\}$ to counter the growth and a gaussian potential $\frac{1}{DV^2}(n_2^{sphere}-V)^2$ to keep the volume approximately constant. If necessary, a gaussian $\frac{1}{DL^2}(q-L)^2$ to fix the boundary is also added. For some experiments, we add a boundary coupling $\exp\{\kappa_b q\}$. When working with the Ising model, Ising spins are placed on the nodes. Each spin interacts with the spins on neighboring nodes with a potential $-JS_iS_j$, except that no spins interact with the special node. Each node on the boundary of the disk interacts with the external field with a potential $-hS_i$. Spins are updated via Metropolis or Wolff cluster algorithms.

We have performed some tests of our code to ensure that it is able to reproduce known analytic results. Warner et. al. [7] hand calculated amplitudes for disks with volumes up to five and boundaries up to seven. For pure gravity we compared these amplitudes to those generated by our code with the following positive results: ($\kappa_2 = \kappa_b = 0$)

V,L	predicted	observed
1,3	1.00	1.00
2,4	1.50	1.50
3,3	1.00	0.99
3,5	3.00	2.99
4,4	3.75	3.75
4,6	7.00	6.96
5,3	3.00	2.99
5,5	12.60	12.60
5,7	18.00	17.93

For the Ising model, we added counting of the spin configurations for $(1,3)$ through $(3,5)$. Modifying the amplitudes for a spin coupling $J = 0.1$, yielded the following:

V, L	predicted	observed
1,3	1.00	1.00
2,4	3.03	3.02
3,3	2.04	2.04
3,5	12.27	12.22

For pure gravity it is known that the partition function should be of the form

$$Z[V, L] = L^a e^{\nu L} V^b e^{\mu V} e^{-L^2/V}$$

with $a = -0.5$ and $b = -2.5$.

Let n_3 be the number of nodes with exactly three surrounding triangles. From the partition function and detailed balance, it can be shown that

$$\ln\{\frac{n_3}{V-2}\} \approx C - 2b/V$$

We generated data to test this and found it was statistically consistent with $b = -2.5$. Because the above partition function is only correct for large volumes, the best results are obtained by excluding disks smaller than some cutoff from the fits. Of course, excluding too much data will harm the fits. See figure 1.

When two sections of the disk share a boundary of only one link, we say that the disk has split into two minimal neck baby universes. Using the partition function, it can be shown that the probability of a disk of volume V and boundary L generating a baby universe of volume V_1 and boundary L_1 is

$$P[V_1, L_1; V, L] =$$
$$\frac{[L_1(L+2-L_1)]^{a+1}}{L^a} * \left[\frac{V_1(V-V_1)}{V}\right]^b * e^{2\nu} *$$
$$\exp\{-\frac{L_1^2}{V_1} - \frac{(L+2-L_1)^2}{V-V_1} + \frac{L^2}{V}\}$$

We generated data to extract a fit of a from this formula and found it not to be inconsistent with $a = -0.5$. Again, the smallest disks are excluded. See figure 2.

The above tests have given us sufficient confidence in our simulations to press on. We have measured magnetization and susceptibility in the bulk of the disk and on the boundary for volumes of 1000 and 2000 with various values of κ_b and J with $h = 0$ and we have the beginnings of a map of the phase diagram along with an idea of what some of its extremes should look like.

We currently plan to check the geometric exponents with an Ising model (expect $a = \frac{1}{3}$ and $b = -\frac{7}{3}$), and map the phase diagram. Afterwards, we will attempt to verify the effects of boundary fields derived by Carroll, Ortiz and Taylor [2]. For instance, they have predicted that the magnetic field on the boundary of the disk should be:

$$\langle m \rangle = \frac{(e^{2h} - 1)(3 + (2 + \sqrt{7})e^{2h})}{(1 + (-1 + \sqrt{7})e^{2h} + (2 + \sqrt{7})e^{4h})}$$

and that the bulk magnetization should be

$$\langle M \rangle = L^{\frac{1}{3}} V^{-\frac{1}{3}}.$$

We then wish to compare these results to those of a three state Pott's model with spins on triangles.

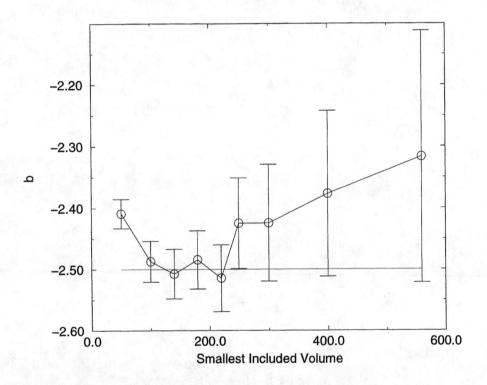

FIGURE 1. b vs. Smallest Included Volume

FIGURE 2. a vs. Smallest Included Boundary

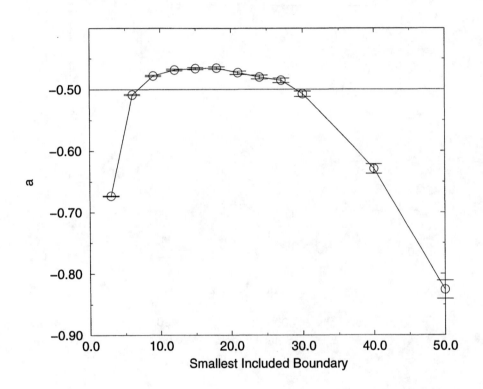

REFERENCES

1. Carroll, Ortiz, and Taylor, hep-th/9605169
2. Carroll, Ortiz, and Taylor, hep-th/9711008
3. Carroll, Ortiz, and Taylor, hep-th/9510199
4. M. Goulian, UCSBTH-91-01
5. J. B. Hartle and S. W. Hawking, Phys. Rev. D28 (1983) 2960
6. Moore, Seiberg, and Staudacher, Nucl. Phys. B362 (1991) 665
7. Warner, Catterall, Renken, hep-lat/9808006
8. Ambjorn, Jurkiewicz, and Watabiki, hep-th/9503108
9. W. T. Tutte, Can. J. Math. 14 (1962) 21
10. F. David, hep-th/9303127

Coherent conversion of neutrino flavour by collisions with relic neutrino gas

I. S. BATKIN AND M. K. SUNDARESAN

Ottawa-Carleton Institute for Physics,
Physics Department, Carleton University,
1125, Colonel By drive, Ottawa, Canada K1S 5B6

Abstract. In this paper, we extend considerations developed for application to coherence in spontaneous radiation processes involving a gas, to interactions of high energy neutrinos with a relic neutrino gas leading to conversion of flavour. Under certain conditions we show that this flavour conversion can be strongly enhanced by coherence effects. These considerations are applied to neutrinos from the Sun and from Supernovae, and estimates for the enhancement effects are given.

INTRODUCTION

Coherent effects play an important role in many physical processes. Early examples of these arise in high energy interference effects in bremsstrahlung and pair production in single crystals. [1]. If l is the characteristic length of the material over which bremsstrahlung or pair production occur, under certain conditions on the momentum transfer at high energies, it is possible that l much larger than the interatomic distance a in the lattice. In such a situation constructive interference of the amplitudes for bremsstrahlung or pair production produced from all the atoms within the distance l occurs. This gives rise to an enhancement factor to the cross section for the process.

That coherent effects can arise in gaseous systems is also well known [2]. When the gas is treated as a single quantum mechanical system, energy levels corresponding to certain correlations between individual molecules of the gas can be obtained. Dicke showed that spontaneous emission of radiation between two such levels can lead to the emission of coherent radiation. Such levels were termed by him as "super-radiant" and the coherent radiation as "super-luminal". He considered the case of a gas of dimension small compared to the wavelength of the radiation emitted as well as the case of a gas of large extent. (In the latter case, the effect of the photon recoil on coherence has to be taken into account). The basic reason why the coherence effect is missing in the usual treatment of spontaneous radiation by a gas is that the molecules of the gas are considered to radiate independently of each other. This is not correct even though it might be argued that as a result of the large distances that separate molecules in a gas, the probability of a given molecule emitting a photon should be independent of the states of other molecules. The usual reasoning overlooks the fact the molecules of the gas are interacting with a common radiation field and hence influence each other and cannot be treated as independent.

In an obvious extension of Dicke's [2] treatment, we could consider coherence effects arising in the scattering reactions of an external particle on the particles of a gas considered as a single quantum mechanical system. In this case the interaction between the external particle and the particles of the gas could be through virtual emission and absorption of radiation, the energy lost or gained by the external particle, manifesting as a transition between the "super-radiant" states of the gas. To illustrate what we have in mind, we consider the interaction of an incident particle with particles of a gas in the form of an isospin-isospin interaction,

$$H_{int} = \sum_j \vec{\tau}.\vec{\tau}_j V_j(\vec{r} - \vec{r}_j),$$

where $\vec{\tau}$ is the isospin of the incident particle, $\vec{r}$ is its coordinate, $\vec{\tau}_j$ is the isospin of the jth particle of the gas, $\vec{r}_j$ is its coordinate, V_j represents an interaction potential between them which might depend on some property at site j, and the sum is over all particles j of the gas. (We work with units in which $\hbar = c = 1$). Let the isospin state of the gas initially be $|I, M>$ and finally be $|I', M'>$. The transition matrix element for an incident particle of momentum $|\vec{p}>$ to a momentum $|\vec{p}'$, will be

$$\mathcal{M} = <\vec{p}'|H_{int}|\vec{p}> = \vec{\tau}. <I', M'|\sum_j \vec{\tau}_j \int d\vec{r} V_j(\vec{r} - \vec{r}_j) \exp(i\vec{q}.\vec{r})|I, M>,$$

where the momentum transfer $\vec{q} = \vec{p} - \vec{p}'$. Now we can write,

$$\int d\vec{r}\, V_j(\vec{r} - \vec{r}_j) \exp i(\vec{q}.\vec{r}) = \exp i(\vec{q}.\vec{r}_j)\, V_j(\vec{q}),$$

and we get

$$\mathcal{M} = \vec{\tau}. <I', M'|\sum_j V_j(\vec{q})\vec{\tau}_j \exp i(\vec{q}.\vec{r}_j)|I, M>.$$

If in the sum over j, the maximum value of $|\vec{r}_j|$ is D (linear dimension of the gas), and $q = |\vec{q}|$ is such that, $qD \ll 1$ then the exponential factor inside the sum can be replaced by unity and the matrix element involves only $\vec{\tau}. <I', M'|\sum_j V_j(\vec{q})\vec{\tau}_j|I, M>$. Suppose $V_j(\vec{q})$ is distributed about some average value $\bar{V}(\vec{q})$ with a rather small spread, we can approximate in the sum over j the value of $V_j(\vec{q})$ by this average value and pull it out of the sum. Then the matrix element involves the factor $\bar{V}(\vec{q})\vec{\tau}. <I', M'|\sum_j \vec{\tau}_j|I, M>$. If the sum $\sum_j \vec{\tau}_j = \vec{\tau}'$, then the matrix element has the factor

$$\tau_z <I', M'|\tau'_z|I, M> + \frac{1}{2}(\tau_+ <I', M'|\tau'_-|I, M> + \tau_- <I', M'|\tau'_+|I, M>,$$

111

where τ'_+, τ'_- are raising and lowering operators for isospin. The important point to note is that under this condition on q

$$< I', M' | \tau'_\pm | I, M > = \delta_{II'} \delta_{M', M\pm 1} \sqrt{(I \mp M)(I \pm M + 1)}.$$

and the matrix element for large I and small M (identified with "super-radiant" state of the gas), is proportional to I and the probability of the scattering with isospin flip is proportional to I^2. For a gas with n particles, $|M| \leq I \leq \frac{1}{2}n$, the probability is proportional to n^2 (coherent process).

The relevance of this model with isospins for our process of flavour conversion will become clear in a later section. Since we could be dealing with particles of extremely small mass, the above model will have to be generalized to be relativistic, in which case the $\vec{q}$ above will become the four-momentum transfer, $q = (q_0, \vec{q})$, with time component, q_0, not necessarily zero. We show below that even in this case, it is possibile to have coherent effects when $q_0 D$ (remember $c = 1$ in our units) and qD are small compared to 1.

Thus coherence effects could appear in this process when the magnitude of three momentum transfer from the external particle is of order $(1/D)$, where D is a scale representing the linear extent of the gas, even though the energy transfer is not zero. It is this extension of Dicke's considerations which we intend to apply in this paper to the flavour conversion of incident neutrinos interacting with relic neutrinos.

According to the standard big-bang cosmological model [3], there exists a sea of relic neutrinos, analogous to the sea of microwave photons. At the decoupling temperature $T_d \simeq 1 MeV$, the neutrinos would still be in equilibrium and the phase-space density of ith flavour of neutrino ν_i and antineutrino $\bar{\nu}_i$ are given by the distribution functions $f_{\nu_i}(k)d\tau$ and $f_{\bar{\nu}_i}(k)d\tau$, where $d\tau = d^3x d^3k$ is the volume element in phase space, and the distribution functions have the equilibrium forms

$$f_{\nu_i}(k) \equiv \frac{1}{\exp[((k^2 + m_i^2)^{1/2} - \mu_{d_i})/T_{d_i}] + 1} \quad (1)$$

and

$$f_{\bar{\nu}_i}(k) \equiv \frac{1}{\exp[((k^2 + m_i^2)^{1/2} + \mu_{d_i})/T_{d_i}] + 1}. \quad (2)$$

Here m_i is the mass of the neutrino of flavour i, and μ_{d_i} is a possible chemical potential included to allow for asymmetry between ν_i and $\bar{\nu}_i$.

In the standard model of cosmology, it is assumed that the neutrino asymmetries are small, so the chemical potential term is zero. Then the number densities for neutrinos and antineutrinos for each flavour are equal, $N_{\nu_i} \simeq N_{\bar{\nu}_i} \simeq 50/cm^3$. Their average momentum is about $5.2 \times 10^{-4} eV$. We will also assume that the chemical potentials for the different flavours are also the same so that their number densities are also equal. Also, we assume that, after decoupling since the particles of the neutrino gas suffer only a cosmological expansion, the relative phases of the particle's wave functions change in a gradual fashion (we do not consider neutrino oscillations here which will only complicate matters further) and therefore it is reasonable to assume that there is a non-random phase relationship between the particles and antiparticles of of the relic neutrino gas at its different sites. This assumption will allow us to define the phase of the amplitude for conversion from different particles of the relic gas.

We consider the following problem. Suppose electron neutrinos are produced in some region of space within a small sphere of radius $D_1, (D_1 \simeq 0)$ (say the solar core or a supernova core). Consider these propagating out from that sphere into outer space and spreading out into a much larger sphere of radius D, the region between the spheres being permeated by relic neutrino gas. The electron neutrinos will annihilate with electron antineutrinos of the relic sea and produce (along with electron neutrinos and electron antineutrinos,) muon neutrinos and muon antineutrinos with a certain probablilty and hence have a certain probability for flavour conversion. If the incident neutrino energies are small compared to the Z^0 mass, this annihilation amplitude can be written in the effective 4-fermion form with the weak Fermi constant G_F, and the probability will be $\propto G_F^2$ and therefore will be extremely small. The question we focus attention on is whether conditions exist under which the relic neutrino gas considered as a single quantum mechanical system, will give rise to a coherent enhancement factor just as in Dicke's work on gaseous molecular systems, and what the size of this enhancement factor is, if it exists.

DESCRIPTION OF CORRELATED STATES OF NEUTRINO GAS

Just as in Dicke's [2] work in dealing with a gas of molecules, in our work in dealing with a gas of relic antineutrinos, it is convenient to introduce a quantity like the isospin to distinguish flavours. Considering only the two flavour case for simplicity in this work, and assuming very small flavour mass differences, we introduce isospin matrices (similar to Pauli matrices) $I_i, i = 1, 2, 3$ and associate the third component $I_3 = +1/2$ with the muon neutrino, ν_μ, and $I_3 = -1/2$ with the electron neutrino, ν_e. For the corresponding antiparticles we then have $I_3 = +1/2$ for the electron antineutrino and $I_3 = -1/2$ for the muon antineutrino. The other components $I_i, i = 1, 2$ can be combined into raising and lowering operators in the standard manner to obtain flavour conversion from one type into the other. The flavour conversions occur, of course, through coupling to the common Z^0 field, through the reaction

$$\nu_e + \bar{\nu}_e \to Z^0 \to \nu_\mu + \bar{\nu}_\mu. \tag{3}$$

Since the incident ν_e converts only on the relic $\bar{\nu}_e$'s we focus attention on the relic antineutrinos only. The Hamiltonian for a relic antineutrino gas of n particles of both flavours, ignoring the Z^0 field, can be written as

$$H = H_0 + E \sum_{j=1}^{n} I_{j3} \qquad (4)$$

where E is the energy difference between the two flavours, and H_0 represents the Hamiltonian for the center of mass motions of the antiparticles of the relic gas. EI_{j3} is the energy of the jth antiparticle and has the eigenvalues $\pm(1/2)E$. Note that H_0 and all the I_{j3} commute with each other. Thus energy eigenfunctions may be chosen to be simultaneous eigenfunctions of $H_0, I_{13}, I_{23}, \cdots, I_{n3}$. We can write a typical energy eigenstate of the relic antineutrino gas as

$$\Psi_{Pm} = \psi_P(x_1, x_2, \cdots, x_n)[+ + - - + + + - \cdots]. \qquad (5)$$

Here the ψ function on the right hand side has as its arguments, the (4-dimensional) coordinates of all the relic antiparticles, P is the total momentum, and the part occurring in the square brackets specifies the flavour content, $+$ for electron antineutrino and $-$ for muon antineutrino respectively. Let n_+, n_- represent the number of antiparticles of the plus and minus type respectively. If there are n_+, plus symbols, and n_-, minus symbols in the square bracket, then the quantum number m for this state is

$$m = \frac{1}{2}(n_+ - n_-), \text{and} \qquad (6)$$

$$n = n_+ + n_- = \text{ total number of relic antiparticles.} \qquad (7)$$

The total energy of this state is

$$E_{Pm} = E_P + mE \qquad (8)$$

The Hamiltonian given in eq.(4) is such that H_0 operates on the antiparticle coordinates only, and gives

$$H_0 \psi_P = E_P \psi_P, \qquad (9)$$

while the I_{j3} operate on the plus or minus symbol in the jth place corresponding to whether the particle is an electron antineutrino or a muon antineutrino. The result of operating with I_{j1} and I_{j2} (in analogy with Pauli matrices) are

$$I_{j1}[\cdots, \pm, \cdots] = \frac{1}{2}[\cdots, \mp, \cdots],$$
$$I_{j2}[\cdots, \pm, \cdots] = \pm \frac{1}{2} i [\cdots, \mp, \cdots] \qquad (10)$$

Let us also define the operators

$$I_k = \sum_{j=1}^{n} I_{jk}, k = 1, 2, 3 \text{ and}$$
$$I^2 = I_1^2 + I_2^2 + I_3^2. \qquad (11)$$

In this notation the Hamiltonian becomes

$$H = H_0 + E I_3 \text{ and}$$
$$I_3 \Psi_{PM} = M \Psi_{PM} \tag{12}$$

To the above we must add the Hamiltonian of the Z^0 field and the interaction between this field and the neutrino gas (as in the standard model) to complete the dynamics.

The operators I_1, I_2, I_3 eq. (11) obey the same commutation relations as the three components of angular momentum. The effective Z^0 interaction operator changes electron type neutrino into muon type (and vice versa) (and a relic electron type antineutrino into a muon type antineutrino) as detailed later. It will thus involve combinations of the operators I_1, I_2, and the transition will obey the selection rule $\Delta M = \pm 1$. The transitions due to the effective interaction mediated by Z^0 will thus connect states of the type given in eq.(5) to states with one fewer plus and one more minus in the square bracket (or one more plus and one fewer minus) only. The methods involved are very closely analogous to methods one uses with isospin 1/2 systems.

Borrowing from the theory of angular momentum, the operators H and I^2 of eq. (11), commute and therefore the energy eigenstates can be simultaneously chosen to be eigenfunctions of I^2. These states are thus linear combinations of states in eq.(5). The operator I^2 has eigenvalues $I(I+1)$ where I is integral or half-odd integral and positive and

$$|M| \leq I \leq \frac{1}{2} n \tag{13}$$

The eigenvalue I may be called the "cooperation number" of the relic antineutrino gas. If we denote these new eigenstates as Ψ_{PIM}, then

$$H \Psi_{PIM} = (E_P + M E) \Psi_{PIM}, \tag{14}$$
$$I^2 \Psi_{PIM} = I(I+1) \Psi_{PIM}. \tag{15}$$

The complete set of eigenstates Ψ_{PIM} may be generated as in the theory of angular momentum, starting with the largest I and the largest M (a unique state) and using lowering operator $I_- = I_1 - i I_2$ successively on the states. This state has $I = M = \frac{1}{2} n$ and is

$$\Psi_{P, \frac{1}{2}n, \frac{1}{2}n} = \psi_P \cdot [+ + + + \cdots + + + +] \tag{16}$$

All the states with this same value of $I = \frac{1}{2} n$ but with lower values of M (are non degenerate also) and can be generated as

$$\Psi_{P,I,M} = [(I^2 - I_3^2 - I_3)^{-1/2} I_-]^{I-M} \Psi_{P,I,I} \tag{17}$$

The fractional power operator in this equation is defined to have positive eigenvalues only. It is included to preserve the normalization. States with the next lower value of $I = \frac{1}{2}n - 1$ can be constructed by first constructing a state orthogonal to $\Psi_{P,I=\frac{1}{2}n,M=\frac{1}{2}n-1}$ and then operating with the lowering operator successively. This can be carried further to generate all the states belonging to other values of I. In this way we get all the states in the new representation.

The current state of the relic gas of antineutrinos is such that nearly equal numbers of electron antineutrinos and muon antineutrinos exist in it. Thus this state has $M \simeq 0$. Expressing it as a linear combination of the new functions $\Psi_{P,I,M\simeq 0}$ for different I, we will assume that the coefficient of the term with large I of order n in the linear combination is not vanishing. This is a resonable assumption in a scenario in which the neutrinos produced initially are of the electron type followed somewhat later by the muon type in more or less equal numbers. This assumption would be invalid in an associated production scenario in which the particles of the two different flavours are produced in pairs. If we consider the current state of the relic gas to be one with large I and small M it will be in a state analogous to the "super radiant" state of the gas molecules of Dicke. Under certain conditions we could expect to see coherence effects in the conversion of the incident neutrino flavour along with an antineutrino in the relic gas changing its flavour at the same time. In flavour conversion transitions the quantum number I does not change while $\Delta M = \pm 1$. We give details of this calculation in the next section.

CALCULATION OF FLAVOUR CONVERSION PROBABILITY IN A RELIC NEUTRINO GAS OF FINITE EXTENT

We first consider annihilation of a specific electron neutrino, labelled by an index i, which can take values $i = 1, 2, 3 \cdots, N_1$. Let us consider a ν_e with 4-momentum $p_{1,i}$, (produced from some location taken as the origin), annihilating on a $\bar{\nu}_e$ of 4-momentum $p_{2,j}$, $j = 1, 2, \cdots, N$, in the relic gas containing a total number N of electron antineutrinos (in a sphere of radius D), mediated in the s channel by Z^0 as in the standard model. We will attribute masses, m_e and m_μ, to the electron neutrino and the muon neutrino and their respective antiparticles. Although not essential for our purpose, we will consider these masses to be so small that their relativistic energies $E(p)$ can be approximated by $p + m^2/(2p)$. We will also assume that $m_e \neq m_\mu$. (The muon antineutrinos in the relic sea are spectators which do not participate in the process of flavour conversion.) The matrix element for the conversion, ν_e to ν_μ, (and hence also of the $\bar{\nu}_e$ to $\bar{\nu}_\mu$ in the relic gas), is

$$\mathcal{M}_i = \sum_{j=1}^{N} \frac{1}{V^2} \frac{-ig^2}{4M_W^2((p_{1,i} + p_{2,j})^2/M_Z^2 - 1)} . F_{ij} .$$
$$(2\pi)^4 \delta_D^4(p_{1,i} + p_{2,j} - p_{3,i} - p_{4,j}) \Pi_{l \neq j} < p_{4,l}|p_{2,l}>, \tag{18}$$

with

$$F_{ij} = \bar{v}(p_{2,j})\gamma^\lambda \frac{1-\gamma_5}{2} u(p_{1,i}) . \bar{u}(p_{3,i})\gamma_\lambda \frac{1-\gamma_5}{2} v(p_{4,j}). \tag{19}$$

Here V is the normalization volume and we normalize all particle wave functions with relativistic normalization of $2E$ particles in volume V and M_W and M_Z are masses of the W and Z bosons. (In eq.(18) the product over $l \neq j$ includes also the muon antineutrino spectators.) Since we are interested in the case where the incident neutrino, produced from some location taken as the origin, and propagates a certain finite distance D in a finite time T ($T = D$) ($\hbar = c = 1$ in our units), we modify the four dimensional δ-function in eq.(18) and write it as

$$(2\pi)^4 \delta_D^4(p_{1,i} + p_{2,j} - p_{3,i} - p_{4,j}) = (2\pi)^3 \delta_D^{(3)}(\vec{Q}_{ij}) \frac{\sin \Omega_{ij} T}{\Omega_{ij}} \tag{20}$$

where

$$\vec{Q}_{ij} = \vec{p}_{1,i} + \vec{p}_{2,j} - \vec{p}_{3,i} - \vec{p}_{4,j} \text{ and} \tag{21}$$
$$\Omega_{ij} = E_{1,i} + E_{2,j} - E_{3,i} - E_{4,j} \tag{22}$$

and δ_D^3 is a function which in the limit $D \to \infty$ is the usual three dimensional delta function. It can be expressed in terms of the spherical Bessel function j_1 as is done later. We consider the case in which $(p_{1,i} + p_{2,j})^2 \ll M_Z^2$ for all i and j, so that the invariant matrix element for the conversion process can be written in the effective four fermion form with $\frac{G_F}{\sqrt{2}} = \frac{g^2}{8M_W^2}$. We further carry out a Fierz transformation of the above so that the factor F in eq.(19) becomes

$$F_{ij} = -\bar{u}(p_{3,i})\gamma_\lambda \frac{(1-\gamma_5)}{2} u(p_{1,i}) . \bar{v}(p_{2,j})\gamma^\lambda \frac{(1-\gamma_5)}{2} v(p_{4,j}). \tag{23}$$

We introduce the four momentum transfer q_i ($q_{0i} = E_{1,i} - E_{3,i}, \vec{q}_i = \vec{p}_{1,i} - \vec{p}_{3,i}$) and we will be particularly interested in the situation $\vec{q}_i \simeq 0$. For small momentum transfer, the effective matrix element leads to the conversion of an incident ν_e of four momentum $p_{1,i}$ to a ν_μ of four momentum $p_{3,i}$ with corresponding changes to the relic antineutrino. Using the isospin formalism described in the previous section to describe this conversion, we attach an isospin factor $< (i)^+, (j)^- | I_{i+} I_{j-} | (i)^-, (j)^+ >$ to M, so that the first factor I_{i+} acts on the first minus symbol on the right (electron neutrino (i)) and changes it into a plus symbol (muon neutrino (i)), while the second factor I_{j-} acts on the relic electron antineutrino (j) with plus symbol on the right and changes it into a minus symbol corresponding to a muon antineutrino (j). All the other spectator electron antineutrinos (those labeled by $l \neq j$) (and also the muon antineutrino spectators) remain unaltered. Thus we get from eq.(18) after putting in these modifications,

$$\mathcal{M}_i = \frac{\sqrt{2} G_F}{V^2} \sum_{j=1}^N F_{ij} . < (i)^+, (j)^- | I_{i+} I_{j-} | (i)^-, (j)^+ >$$
$$\Pi_{l \neq j} < p_{4,l} | p_{2,l} > < (l)^+ | (l)^+ > < (l)^- | (l)^- >$$
$$(2\pi)^3 \delta_D^{(3)}(\vec{Q}_{ij}) . \frac{\sin \Omega_{ij} T}{\Omega_{ij}} \tag{24}$$

where again the spectator muon antineutrinos (minus symbols) are explicitly shown in the product over $l \neq j$. As mentioned in the last section, we go over to the (I, M) basis for the isospin functions to introduce the correlations between the antineutrinos in the relic sea. As there are N electron antineutrinos and N muon antineutrinos, the maximum value of I is N and we have

$$|M| \leq I \leq N \qquad (25)$$

The probability of flavour conversion by the above reaction for the incident electron neutrino labelled by i is obtained by finding $|\mathcal{M}_i|^2$ and integrating it over the final phase space of all the particles of the relic gas. $\mathcal{M}_i$ involves sum over all the electron antineutrinos labelled by j, and will contain in addition to direct terms, terms which represent interference effects between amplitudes form different electron antineutrinos j and k. The direct terms will give the usual incoherent probability (proportional to the number N of electron antineutrinos in the relic gas). We show that when the momentum transfer is such that $|\vec{q}| \simeq (1/D)$ and $q_0 \simeq (1/T)$, coherent effects proportional to the square of the number of particles contained in the sphere of radius R arise in the expression for the probability. The expression for the probability is

$$dP_i = |\mathcal{M}_i|^2 \frac{V d^3 p_{3,i}}{(2\pi)^3 2 E_{3,i}} \prod_{j=1}^{2N} \frac{V d^3 p_{4,j}}{(2\pi)^3 2 E_{4,j}} \qquad (26)$$

For $|\mathcal{M}_i|^2$ we get

$$|\mathcal{M}_i|^2 = \frac{2 G_F^2}{V^4} \left[\sum_{j=1}^{N} |F_{ij}|^2 | <(i)^+|I_{i+}|(i)^-> |^2 | <(j)^-|I_{j-}|(j)^+> |^2 \right.$$

$$\prod_{l \neq j} | <p_{4,l}|p_{2,l}> |^2 | <(l)^+|(l)^+> |^2 | <(l)^-|(l)^-> |^2$$

$$\left((2\pi)^3 \delta_D^{(3)}(\vec{Q}_{ij}) \frac{\sin \Omega_{ij} T}{\Omega_{ij}} \right)^2$$

$$+ \sum_{j=1}^{N} \sum_{k=1}^{N} [F_{ij} <p_{4,k}|p_{2,k}> <(j)^-|I_{j-}|(j)^+> <(k)^+|(k)^+> .$$

$$F_{ik}^* <p_{4,j}|p_{2,j}>^* <(k)^-|I_{k-}|(k)^+>^* <(j)^+|(j)^+>^*$$

$$\left((2\pi)^3 \delta_D^{(3)}(\vec{Q}_{ij}) \frac{\sin \Omega_{ij} T}{\Omega_{ij}} (2\pi)^3 \delta_D^{(3)}(\vec{Q}_{ik}) \frac{\sin \Omega_{ik} T}{\Omega_{ik}} \right)$$

$$\prod_{l \neq j,k} | <(l)^+|(l)^+> |^2 | <(l)^-|(l)^-> |^2 | <p_{4,l}|p_{2,l}> |^2$$

$$+ \text{complex conjugate}]] \qquad (27)$$

To evaluate this further, we need the F_{ij}. In the limit of small masses for the neutrinos, the dominant contribution comes only when the neutrinos are left handed and the antineutrinos are right handed. Evaluating this using helicity states, we get

$$F_{ij} = 4\sqrt{p_{1i}p_{4j}(1-\cos\Theta_{1i,4j})}\sqrt{p_{3i}p_{2j}(1-\cos\Theta_{3i,2j})}\exp i(\Phi_{3i,2j} - \Phi_{1i,4j}) \quad (28)$$

where the various $p's$ occurring within the square roots are magnitudes of the three momenta, the 3-vectors themselves being specified in spherical basis, $\vec{p}_{1,i} = (p_{1i}, \theta_{1i}, \phi_{1i}), \vec{p}_{2,j} = (p_{2j}, \theta_{2j}, \phi_{2j}), \vec{p}_{3,i} = (p_{3i}, \theta_{3i}, \phi_{3i})$, and $\vec{p}_{4,j} = (p_{4j}, \theta_{4j}, \phi_{4j})$. Also, the $\Theta's$ and $\Phi's$ occurring in eq.(??) are

$$\Theta_{1i,4j} = \cos\theta_{1i}\cos\theta_{4j} + \sin\theta_{1i}\sin\theta_{4j}\cos(\phi_{1i} - \phi_{4j}) \quad (29)$$

$$\Theta_{3i,2j} = \cos\theta_{3i}\cos\theta_{2j} + \sin\theta_{3i}\sin\theta_{2j}\cos(\phi_{3i} - \phi_{2j}) \quad (30)$$

$$\Phi_{1i,4j} = \arctan\left(\frac{\sin(\phi_{4j} - \phi_{1i})}{\cot(\theta_{1i}/2)\tan(\theta_{4j}/2) - \cos(\phi_{4j} - \phi_{1i})}\right) \quad (31)$$

$$\Phi_{3i,2j} = \arctan\left(\frac{\sin(\phi_{2j} - \phi_{3i})}{\cot(\theta_{3i}/2)\tan(\theta_{2j}/2) - \cos(\phi_{2j} - \phi_{3i})}\right) \quad (32)$$

Let us call the contribution from the first term of eq.(27) as the incoherent contribution to dP_i and refer to it as dP^i_{incoh} and the contribution from the second term of eq.(27) as the coherent contribution and refer to it as dP^i_{coh}. We now evaluate each in turn.

In evaluating the first term, we can work in the limit when $D, T \to \infty$ which allows one to replace the product of the *sine* function and the delta function in eq.(27), by the 4-dimensional energy momentum conserving delta function for the $2 \to 2$ particle process. Thus P^i_{incoh} is (we suppress writing isospin factors for the spectators which give 1)

$$P^i_{incoh} = \frac{2G_F^2}{V^4}\sum_{j=1}^N |F_{ij}|^2 \cdot |<(i)^+|I_{i+}|(i)^->|^2|<(j)^-|I_{j-}|(j)^+>|^2$$

$$[(2\pi)^4\delta^{(4)}(p_{1,i} + p_{2,j} - p_{3,i} - p_{4,j})]^2 \frac{Vd^3p_{3,i}}{(2\pi)^3 2E_{3,i}}\frac{Vd^3p_{4,j}}{(2\pi)^3 2E_{4,j}}$$

$$\prod_{(l\neq j)=1}^N |<p_{4,l}|p_{2,l}>|^2 \frac{Vd^3p_{4,l}}{(2\pi)^3 2E_{4,l}} \quad (33)$$

With our normalization we have

$$|<p_{4,l}|p_{2,l}>|^2 = [(2\pi)^3\frac{2E_{4,l}}{V}\delta^3(\vec{p}_{2,l} - \vec{p}_{4,l})]^2. \quad (34)$$

Thus, we get

$$\prod_{l\neq j=1}^{2N}\int \frac{Vd^3p_{4,l}}{(2\pi)^3 2E_{4,l}}|<p_{4,l}|p_{2,l}>|^2 = \prod_{l\neq j=1}^{2N}\frac{(2\pi)^3\delta^3(0)}{V}2E_{2,l}$$

$$= \prod_{l\neq j=1}^{2N} 2E_{2,l} \quad (35)$$

since $(2\pi)^3\delta^3(0) = V$ and $E_{2,l} = E_{4,l}$ as the momenta are equal. Thus we have

$$P^i_{incoh} = \frac{2G_F^2}{V^2}[(2\pi)^4\delta^4(0)]\sum_j^N \prod_{l\neq j=1}^{2N} 2E_{4,l} \int \frac{d^3p_{3,i}}{(2\pi)^3 2E_{3,i}} \int \frac{d^3p_{4,j}}{(2\pi)^3 2E_{4,j}}$$
$$(2\pi)^4\delta^{(4)}(p_{1,i} + p_{2,j} - p_{3,i} - p_{4,j})$$
$$|F_{ij}|^2| < (i)^+|I_{i+}|(i)^- > |^2| < (j)^-|I_{j-}|(j)^+ > |^2 \quad (36)$$

Now we note from eq.(28) that $|F_{ij}|^2$ involves $(p_{1,i}.p_{4,j})(p_{3,i}.p_{2,j})$, (all the p's here 4-vectors), we introduce

$$B_{ij} = \int \frac{d^3p_{3,i}}{(2\pi)^3 2E_{3,i}} \int \frac{d^3p_{4,j}}{(2\pi)^3 2E_{4,j}}(2\pi)^4\delta^{(4)}(p_{1,i} + p_{2,j} - p_{3,i} - p_{4,j})(p_{1,i}.p_{4,j})(p_{3,i}.p_{2,j}) \quad (37)$$

which is a well known integral over 2-body phase space. When evaluated it is (using $X_{ij}^\alpha = p_{1,i}^\alpha + p_{2,j}^\alpha = p_{3,i}^\alpha + p_{4,j}^\alpha$),

$$B_{ij} \simeq \frac{X_{ij}^2}{96\pi}\left[(p_{1,i}.p_{2,j}) + \frac{2(p_{1,i}.X_{ij})(p_{2,j}.X_{ij})}{X_{ij}^2}\right] \quad (38)$$

Then P^i_{incoh} becomes

$$\frac{(2\pi)^4\delta^{(4)}(0)}{V^2}\sum_{j=1}^N | < (i)^+|I_{i+}|(i)^- > |^2| < (j)^-|I_{j-}|(j)^+ > |^2$$
$$32G_F^2 B_{ij} \prod_{l\neq j=1}^{2N} 2E_{2,l} \quad (39)$$

Now we average $B_{ij}\prod_{l\neq j=1}^{2N} 2E_{2,l}$ over the distribution of the relic electron antineutrinos with the distribution functions given in eq.(2), and call it $\bar{B}\prod_{l\neq j=1}^{2N} 2\bar{E}_{2,l}$ and pull it outside the sum. Then the sum involves only

$$\sum_{j=1}^N < (j)^-|I_{j-}|(j)^+ > |^2 = N \quad (40)$$

and we get

$$P^i_{incoh} = \frac{(2\pi)^4\delta^{(4)}(0)}{V^2}\prod_{l\neq j=1}^N 2\bar{E}_{2,l}.32G_F^2\bar{B}N| < (i)^+|I_{i+}|(i)^- > |^2 \quad (41)$$

This is the probability for conversion in volume V, when $2E_{1,i}$ particles in the volume are incident on $2E_{2,j}, j = 1, 2, \cdots, 2N$ target particles in the volume. Thus the probability per incident particle on the gas in the target is obtained by dividing the above P_{incoh} by $2E_{1,i}\prod_{j=1}^{2N} 2\bar{E}_{2,j}$ and we get

$$\mathcal{P}^i_{incoh} = \frac{(2\pi)^4 \delta^{(4)}(0)}{V^2} \frac{32 G_F^2 \bar{B} N}{2 E_{1,i} 2 \bar{E}_2} |<(i)^+|I_{i+}|(i)^->|^2 \qquad (42)$$

We can write this as

$$\mathcal{P}^i_{incoh} = D.n_0.\sigma |<(i)+|R_{i+}|(i)->|^2 \qquad (43)$$

where $n_0 = \frac{N}{V}$, $D = cT, (c=1)$ and $\sigma_i = \frac{8 G_F^2 \bar{B}}{E_{1,i} \bar{E}_2}$.

We may perform an average over the intial particle distribution of momenta, which amounts to replacing $E_{1,i}$ by $\bar{E}_1$, and we can perform the sum over i. This will give us a factor N_1 and we will have

$$\mathcal{P}_{incoh} = D.n_0.\sigma N_1 \qquad (44)$$

where σ is $\sigma = \frac{8 G_F^2 \bar{B}}{\bar{E}_1 \bar{E}_2}$.

Now we proceed to the evaluation of the contribution of the second term in the probability of conversion, namely, the coherent contribution. We have (again we suppress isospin factors for spectators which work out to 1)

$$\begin{aligned}
P^i_{coh} = \frac{2 G_F^2}{V^4} \sum_j \sum_k \Big\{ & \int \frac{V d^3 p_{4,j}}{(2\pi)^3 2 E_{4,j}} \cdot \int \frac{V d^3 p_{4,k}}{(2\pi)^3 2 E_{4,k}} |<(i)^+|I_{i+}|(i)^->|^2 \\
& (2\pi)^3 \frac{2 E_{4,k}}{V} \delta_D^{(3)}(\vec{p}_{2,k} - \vec{p}_{4,k}) F_{ij} (2\pi)^4 \delta_D^{(4)}(p_{1,i} + p_{2,j} - p_{3,i} - p_{4,j}) \\
& (2\pi)^3 \frac{2 E_{4,j}}{V} \delta_D^{(3)}(\vec{p}_{2,j} - \vec{p}_{4,j}) F^*_{ik} (2\pi)^4 \delta^{(4)}(p_{1,i} + p_{2,k} - p_{3,i} - p_{4,k}) \\
& <(j)^-|I_{j-}|(j)^+><(k)^-|I_{k-}|(k)^+>^* \\
& +\text{complex conjugate} \Big\} \frac{V d^3 p_{3i}}{(2\pi)^3 2 E_{3i}} \cdot \prod_{l \neq j,k} 2 E_{2,l} \qquad (45)
\end{aligned}$$

Before going into the evaluation of P^i_{coh}, we examine phases. The δ_D^4 functions arise from phase evolutions:

$$\exp i\vec{Q}_{ij}.\vec{r} \exp i\Omega_{ij}(t - t_{2j}) \times \exp(-i\vec{Q}_{ik}.\vec{r}) \exp(-i\Omega_{ik}(t - t_{2k})),$$

when integrated on $\vec{r}$ over a finite region D and on t over a finite region T. Interference can arise only if these phase factors are of order 1. Considering the time part,

$$\exp i(\Omega_{ij} - \Omega_{ik})t \exp i(\Omega_{ik} t_{2k} - \Omega_{ij} t_{2j}),$$

the first factor is of order 1 if $|\Omega_{ij} - \Omega_{ik}| \leq \frac{\pi}{2} T$, and using this information in the second factor, we must have $\Omega_{ij}(t_{2k} - t_{2j}) = \Omega_{ij}(\frac{|\vec{r}_{jk}|^{max}}{c=1}) \leq \frac{\pi}{2}$. Then all the exponentials will be positive for all j and k. This condition implies that $\Omega_{ij} \leq \frac{\pi}{2T}$ for all i and j. Since

$$\Omega_{ij} = E_{1i} - E_{3i} + E_{2j} - E_{2k}$$
$$\simeq \frac{\Delta m^2}{2p_{1i}} + \frac{\Delta m^2}{2p_{2j}}, \qquad (46)$$

where $\Delta m^2 = m_e^2 - m_\mu^2$, it implies that since p_{1i} is large, ignoring the term involving it, the condition for coherence is

$$\frac{\Delta m^2}{2\bar{p}_2} \leq \frac{\pi}{2D}, \qquad (47)$$

where $\bar{p}_2$ is the average momentum of the relic particles. Thus the minimum value q_0 can take depends on Δm^2 which is restricted to $\pi \bar{p}_2/D$.

To evaluate P^i_{coh} we split up the two 4-dimensional delta functions into products of 3-dimensional delta function (momentum conserving) and 1-dimesional delta function (energy conserving) respectively. Integrating over $\vec{p}_{4,j}$ and $\vec{p}_{4,k}$ we must set $\vec{p}_{4,j} = \vec{q}_i + \vec{p}_{2,j}$ and $\vec{p}_{4,k} = \vec{q}_i + \vec{p}_{2,k}$ respectively, upto terms of order $(1/D)$ in the integrand. The 1-dimensional delta functions become $\delta(E_{1,i}+E_{2,j}-E_{3,i}-E_{4,j}(|\vec{p}_{2,j}+\vec{q}_i|))$ and $\delta(E_{1,i}+E_{2,k}-E_{3,i}-E_{4,k}(|\vec{p}_{2,k}+\vec{q}_i|))$. The two other 3-dimensional delta functions become each $\delta_D^{(3)}(\vec{q}_i)$. These imply, $|\vec{q}_i|D < \pi/2$. Since the vectors $\vec{p}_{4j}$ and $\vec{p}_{4k}$ are equal to the vectors $\vec{p}_{2j}$ and $\vec{p}_{2k}$ respectively, the Φ dependent phase factors in $F_{ij}F^*_{ik}$ are such that, those exponentials are also nearly unity.

Taking all these into account, we get

$$P^i_{coh} = \frac{2G_F^2}{V^4} \sum_j \sum_k \left\{ \int \frac{V d^3 p_{3,i}}{(2\pi)^3 2E_{3,i}} [(2\pi)^3 \delta_D^{(3)}(\vec{p}_{1,i} - \vec{p}_{3,i})]^2 \cdot | <(i)^+|I_{i+}|(i)^- >|^2 \right.$$
$$[32 \, p_{1i}p_{2j}(1 - \cos\Theta_{1i,2j})p_{3i}p_{2k}(1 - \cos\Theta_{3i,2k})$$
$$<(j)^-|R_{j-}|(j)^+><(k)^-|R_{k-}|(k)^+>^*$$
$$+\text{complex conjugate}]$$
$$\left. \frac{\sin\Omega_{ij}T}{\Omega_{ij}} \frac{\sin\Omega_{ik}T}{\Omega_{ik}} \right\} \prod_{l \neq j,k} 2E_{2,l} \qquad (48)$$

We average this over the relic particle momentum p_2 distribution, so that the p_{2j}'s get replaced by $\bar{p}_2$ and the $\cos\Theta$ factors average to zero. Finally the probability per incident neutrino on the relic gas in volume V can be written as

$$\mathcal{P}^i_{coh} = \frac{64G_F^2 C_i}{V^2} | <(i)^+|I_{i+}|(i)^- >|^2$$
$$\sum_j \sum_k <(j)^-|I_{j-}|(j)^+><(k)^-|I_{k-}|(k)^+>^* \qquad (49)$$

where C_i is the integral

$$C_i = \int \frac{d^3 q_i}{(2\pi)^3} \frac{4\pi D^2}{|\vec{q}_i|} j_1(|\vec{q}_i|D) \left(\frac{\sin^2(p_{1i} - |\vec{p}_{1i} - \vec{q}_i| - \Delta)D}{(p_{1i} - |\vec{p}_{1i} - \vec{q}_i| - \Delta)D)^2} \right) \qquad (50)$$

where the abbreviation Δ has been used for $\Delta m^2/(2\bar{p}_2)$. This integral C_i is delicate to evaluate; it clearly depends on the sign of Δ, that is, it depends on the squared mass difference between the electron and muon neutrinos and thus gives a direction for the probability of conversion. Note that the integral C_i has the dimension of squared length $\simeq D^2$. The double sum over j and k give a factor N^2 which is the coherence factor amplifying this term. Thus finally we have

$$\mathcal{P}^i_{coh} = \frac{64 G_F^2 C_i}{V^2} |<(i)^+|I_{i+}|(i)^->|^2 N^2 \tag{51}$$

Now we may pay attention to the index i which referred to the incident neutrino with momentum p_{1i}. Just as we averaged over the relic momentum distribution we can averge the above over the incident momentum distribution also which will have the effect of replacing the p_{1i}'s by its average, $\bar{p}_1$. Then the sum over i can be carried out and will give rise to a further factor of N_1. Thus finally we have

$$\mathcal{P}_{coh} = \frac{64 G_F^2 \bar{C}}{V^2} N_1 N^2 \tag{52}$$

CONCLUSION

We have shown that if the mass differences of the electron and the muon neutrinos are sufficiently small (smaller than the average relic neutrino energies), the probability of flavour conversion of the incident neutrino acquires a coherence factor of N^2, where N is the number of relic gas particles in the volume of the gas sphere of radius D. If the number of incident particles in the volume is N_1, then we get this factor appearing in the probability. If we extended interference considerations also to the incident particles, there is another potential factor of N_1. Thus, under our assumptions on the formation of the relic gas from the initial big bang, it is possible to overcome the small values due to the smallness of G_F^2 by factors $N_1^2 N^2$ which can be large. If the assumptions under which our calculations are done are valid, the problems of neutrino oscillations (which we have not considered in this work) will become intertwined with the problem of flavour conversion due to the relic antineutrinos.

REFERENCES

1. Dyson F.J., and Uberall H., *Physical Review* **99**, 604,(1955); Uberall H., *Physical Review* **103**, 1055, (1956).
2. Dicke R. H., *Physical Review* **93**, 99, (1954).
3. Weinberg S., *Gravitation and Cosmology*, New York: Wiley 1972.

Local and Global Duality and the Determination of $\alpha(M_Z)$

Stefan Groote

Institut für Physik, Johannes-Gutenberg-Universität, Staudinger Weg 7, 55099 Mainz, Germany
Floyd R. Newman Laboratory of Nuclear Studies, Cornell University, Ithaca, NY 14853, USA

Abstract. This talk presents work concepts and results for the determination of the fine structure constant α at the Z_0 mass resonance. The problem consisting of the break-down of global duality for singular integral weights is circumvented by using a polynomial fit which mimics this weight function. This method is conservative in the sense that it is mostly independent of special assumptions. In this context the difference between local and global duality is explained.

INTRODUCTION

There is a great deal of interest in the accurate determination of the running fine structure constant α at the scale of the Z_0 mass [1–4]. The value of $\alpha(M_Z)$ is of paramount importance for all precision tests of the Standard Model. Furthermore, an accurate knowledge of $\alpha(M_Z)$ is instrumental in narrowing down the the mass window for the last missing particle of the Standard Model, the Higgs particle.

The main source of uncertainty in the determination of $\alpha(M_Z)$ is the hadronic contribution to the e^+e^- annihilations needed for this evaluation. The necessary dispersion integral that enters this calculation has in the past been evaluated by using experimental e^+e^- annihilation data. Discrepancies in the experimental data between different experiments suggest large systematic uncertainties in each of the experiments. In order to reduce the influence of the systematic uncertainties on the determination of $\alpha(M_Z)$ one may attempt to add some theoretical input to the evaluation of the hadronic contribution to $\alpha(M_Z)$.

M. Davier and A. Höcker [3] use QCD perturbation theory in form of local duality (explained later on) in the region above $s = (1.8\,\text{GeV})^2$ for the light flavours, while J.H. Kühn and M. Steinhauser [5] use perturbative results for energy regions outside the charm and bottom threshold regions. Our approach [6] is quite different. We attempt to minimize the influence of data in the dispersion integral over the whole energy region including the threshold regions.

Local and global duality

For the e^+e^- annihilation process there is a main connection between the spectral density $\rho(s)$ and the two-point correlator $\Pi(q^2)$ given in kind of the dispersion relation

$$\Pi(q^2) = \int_{s_0}^{\infty} \frac{\rho(s)ds}{s+q^2} \qquad (1)$$

which implies the reverse relation

$$\rho(s) = \frac{1}{2\pi i} Disc\, \Pi(s) \qquad (2)$$

where the discontinuity is given by

$$Disc\, \Pi(q^2) = \Pi(q^2 e^{-i\pi}) - \Pi(q^2 e^{i\pi}) \qquad (3)$$

and $s_0 = 4m_\pi^2$ is the production threshold of the light flavours. These relations can be a chain between the theory, i.e. the two-point correlator function within perturbative QCD on the one hand and the experiment, i.e. the spectral density or, equivalently, the total cross section on the other hand. But there is one obstacle in using these relations: As depending on methods of functional analysis, the inverse of the dispersion relation is only valid if there are no poles encircled by the path in the complex plane, a condition which is necessary to obtain this relation. These poles can have their origin from weight functions in combination with the spectral density. This means that if there is such a weight function included in the integration of the spectral density, the inverse relation shown above is only valid *locally* and not *globally*. We call this *local* resp. *global duality*.

The experiment side

The hadronic contribution to $\alpha(M_Z)$ which we are concentrating on is given by the integral [1]

$$\Delta\alpha_{\text{had}}(M_Z) = \frac{\alpha}{3\pi} Re \int_{s_0}^{\infty} R(s)H(s)ds \qquad (4)$$

where $R(s)$ is the total e^+e^- hadronic cross section and $H(s)$ is the weight function

$$H(s) = \frac{M_Z^2}{s(M_Z^2 - s)}. \qquad (5)$$

The hadronic cross section is related to the spectral density by

$$R(s) = 12\pi^2 \rho(s). \qquad (6)$$

But we see: the weight function $H(s)$ is indeed singular at the points $s = 0$ and $s = M_Z^2$ on the real axis. So global duality is not valid any more.

The theory side

The two-point correlator is given by

$$i \int \langle 0|j_\alpha^{\text{em}}(x) j_\beta^{\text{em}}(0)|0\rangle e^{iqx} d^4x = (-g_{\alpha\beta}q^2 + q_\alpha q_\beta)\Pi(q^2) \tag{7}$$

where we only included the isospin contribution $I = 1$, in contrast to corresponding considerations for the τ decay. The scalar correlator function $\Pi(q^2)$ consists of perturbative and non-perturbative contributions which we include to the extend we need them to keep the accuracy. For the perturbative contribution to the correlator we use a result given in ref. [7]. I only write down the first few terms,

$$\Pi^P(q^2) = \frac{3}{16\pi^2} \sum_{i=1}^{n_f} Q_i^2 \left[\frac{20}{9} + \frac{4}{3}L + C_F \left(\frac{55}{12} - 4\zeta(3) + L\right) \frac{\alpha_s}{\pi} + O(\alpha_s^2, m_q^2/q^2) \right] \tag{8}$$

with $L = \ln(\mu^2/q^2)$ while in ref. [7] the expression is given up to $O(\alpha_s^2, m_q^{12}/q^{12})$. The number of active flavours is denoted by n_f. For the zeroth order term in the m_q^2/q^2 expansion we have added higher order terms in α_s,

$$\frac{3}{16\pi^2} \sum_{i=1}^{n_f} Q_i^2 \left[\left(c_3 + 3k_2 L + \frac{1}{2}(k_0\beta_1 + 2k_1\beta_0)L^2\right) \left(\frac{\alpha_s}{\pi}\right)^3 + O(\alpha_s^4) \right] \tag{9}$$

with $k_0 = 1$, $k_1 = 1.63982$ and $k_2 = 6.37101$. We have denoted the yet unknown constant term in the four-loop contribution by c_3. Remark, however, that the constant non-logarithmic terms will not contribute to our calculations. The non-perturbative contributions are given in ref. [8],

$$\Pi^{NP}(q^2) = \frac{1}{18q^4} \left(1 + \frac{7\alpha_s}{6\pi}\right) \langle \frac{\alpha_s}{\pi} G^2 \rangle$$
$$+ \frac{8}{9q^4} \left(1 + \frac{\alpha_s}{4\pi} C_F + \ldots \right) \langle m_u \bar{u}u \rangle + \frac{2}{9q^4} \left(1 + \frac{\alpha_s}{4\pi} C_F + \ldots \right) \langle m_d \bar{d}d \rangle$$
$$+ \frac{2}{9q^4} \left(1 + \frac{\alpha_s}{4\pi} C_F + (5.8 + 0.92L)\frac{\alpha_s^2}{\pi^2}\right) \langle m_s \bar{s}s \rangle$$
$$+ \frac{\alpha_s^2}{9\pi^2 q^4} (0.6 + 0.333L) \langle m_u \bar{u}u + m_d \bar{d}d \rangle$$
$$- \frac{C_A m_s^4}{36\pi^2 q^4} \left(1 + 2L + (0.7 + 7.333L + 4L^2)\frac{\alpha_s}{\pi}\right) \tag{10}$$
$$- \frac{448\pi}{243 q^6} \alpha_s |\langle \bar{q}q \rangle|^2 + O(q^{-8})$$

where we have included the m_s^4/q^4-contribution arising from the unit operator. In this expression we used the $SU(3)$ colour factors $C_F = 4/3$, $C_A = 3$, and $T_F = 1/2$. For the coupling constant α_s as well as for the running quark mass we use four-loop expression given in refs. [9–11] even though in both cases the three-loop accuracy would already have been sufficient for the present application.

INTRODUCING OUR METHOD

Our method is based on the fact that we can use global duality when the weight function is non-singular. This is the case for a polynomial function. So we mimic the weight function by a polynomial function obeying different conditions which we will explain later. By adding and subtracting this polynomial function $P_N(s)$ of given order N to the weight function $H(s)$, we obtain without any restrictions

$$\int_{s_a}^{s_b} \rho(s)H(s)ds = \int_{s_a}^{s_b} \rho(s)\left(H(s) - P_N(s)\right)ds + \int_{s_a}^{s_b} \rho(s)P_N(s)ds \qquad (11)$$

where $[s_a, s_b]$ is any interval out of the total integration range. But because the second term has now a polynomial weight, we can use global duality to write

$$\int_{s_a}^{s_b} \rho(s)P_N(s)ds = \frac{1}{2\pi i}\int_{s_a}^{s_b} \text{Disc}\,\Pi(s)P_N(s)ds =$$
$$= -\frac{1}{2\pi i}\oint_{|s|=s_a} \Pi(-s)P_N(s)ds + \frac{1}{2\pi i}\oint_{|s|=s_b} \Pi(-s)P_N(s)ds. \qquad (12)$$

Thus this part can be represented by a difference of two circle integrals in the complex plane. On the other hand, the difference $H(s) - P_N(s)$ suppresses the contribution of the first part. Our method consists thus of the following steps:

- replacing $\rho(s)$ in the first part of Eq. (11) by the value of the experimentally measured total cross section $R(s)$ (see e.g. ref. [1])

- replacing the circle integral contribution to flavours at their threshold by zero

- in all other cases inserting the QCD perturbative and non-perturbative parts of $\Pi(-s)$ on the circle

These replacements can be seen as a concept within QCD sum rules. To obtain the best efficiency of our method, we have to restrict the polynomial function by the following contraints:

- The method of least squares should be used to mimic the weight

- However, the degree N should not be higher than the order of the highest perturbative resp. non-perturbative contribution increased by one (this is a consequence of the Cauchy's theorem which is involved in the analytical integration of the circle integrals)

- Especially for the low energy region, the polynomial function should vanish on the real axis to avoid instanton effects

- In regions where resonances occur, the polynomial function should fit the weight function to suppress those contributions which constitute the highest uncertainty of the experimental data

As just mentioned, the integration on the circle can be done analytically by using the Cauchy's theorem. But we have to keep in mind that the result for $\Pi(-s)$ we use here depends logarithmically on the renormalization scale μ and on the parameters of the theory that are renormalized at the scale μ. These are the strong coupling constant, the quark masses and the condensates. As advocated in [12], we implement the renormalization group improvement for the moments of the electromagnetic correlator by performing the integrations over the circle with radius $|s| = s_b$ with constant parameters, i.e. they are renormalized at a fixed scale μ. Subsequently these parameters are evolved from this scale to $\mu^2 = s_a$ using the four-loop β function. In other words, we impose the renormalization group equation on the moments rather than on the correlator itself. This procedure is not only technically simpler but also avoids possible inconsistencies inherent to the usual approach where one applies the renormalization group to the correlator, expands in powers of $\ln(s/\mu^2)$ and carries out the integration in the complex plane only at the end. In the present case the reference scale is given by $\Lambda_{\overline{\text{MS}}}$.

Subdividing the integration interval

As a first interval we select the range from the light flavour production threshold $s_0 = 4m_\pi^2$ and the next threshold marked by the mass of the ψ, $s_1 = m_\psi^2 \approx (3.1\,\text{GeV})^2$. In this case we set the inner circle integral to zero and obtain

$$\int_{s_0}^{s_1} R(s) H(s) ds = \int_{s_0}^{s_1} R^{\text{exp}}(s) \left(H(s) - P_N(s)\right) ds + 6\pi i \oint_{|s|=s_1} \Pi^{\text{QCD}}(-s) P_N(s) ds. \tag{13}$$

As mentioned above, we impose the constraints to the polynomial function that it should vanish on the real axis at $s = s_1$ and should coincide with the weight function at the ρ resonance, i.e. for $s = m_\rho^2 \approx (1\,\text{GeV})^2$. Fig. 1 shows polynomials of different order in comparison with the weight function. The results shown in Fig. 2 are compared with the result obtained by using only the experimental data. For the up and down quarks we only keep the mass independent part of the QCD contribution while for the strange quark we include also the terms to order $O(m_s^2/q^2)$.

The second interval is limited by s_1 and the threshold marked by the mass of the Υ, $s_2 = m_\Upsilon^2 \approx (9.46\,\text{GeV})^2$. For the charm quark, we again set the inner circle integral to zero, but for the lighter quarks we have to keep both. The perturbative series for the charm quark is used up to it's known extend.

The third interval given between s_2 and $(40\,\text{GeV})^2$ is again subdivided into two pieces because of it's length. For the first of these two intervals we choose $[(9.46\,\text{GeV})^2, (30\,\text{GeV})^2]$, for the second $[(30\,\text{GeV})^2, (40\,\text{GeV})^2]$. Now the bottom quark is the one for which the "threshold rule" (i.e. leaving out the inner circle) applies. The remaining part of the integral starting from $s_4 = (40\,\text{GeV})^2$ up to infinity is done using local duality, i.e. by inserting the function $R(s)$ obtained for perturbative QCD into the second part of Eq. (11).

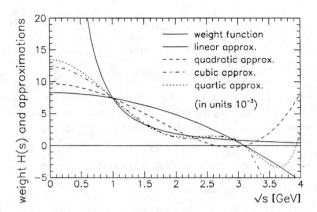

FIGURE 1. Weight function $H(s)$ and polynomial approximations $P_N(s)$ in the lowest energy interval $2m_\pi \leq \sqrt{s} \leq 3.1\,\text{GeV}$. The least square fit was done in the interval $m_\rho \leq \sqrt{s} \leq 3.1\,\text{GeV}$ with further constraints $H(s) = P_N(s)$ at $\sqrt{s} = 1\,\text{GeV}$ and $P_N(s) = 0$ at $\sqrt{s} = 3.1\,\text{GeV}$. The quality of the polynomial approximations are shown up to $N = 4$. We use the scaled variable s/s_1 for the polynomial approximation where s_1 is the upper radius such that $P_N(s/s_1)$ is dimensionless.

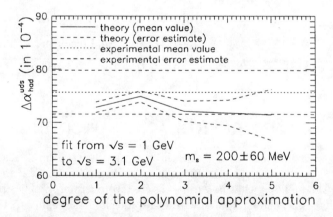

FIGURE 2. Comparison of the l.h.s. and r.h.s. of the sum rule given by Eq. (13) in the interval $0.28\,\text{GeV} \leq \sqrt{s} \leq 3.1\,\text{GeV}$. Dotted horizontal line: value of integrating the l.h.s. using experimental data including error bars [1]. The points give the values of the r.h.s. integration for various orders N of the polynomial approximation. Straight line interpolations between the points are for illustration only. The dashed lines indicate the error estimate of our calculation.

Our results are collected in Table 1. To obtain these results, we used the condensate values

$$\langle \frac{\alpha_s}{\pi} GG \rangle = (0.04 \pm 0.04)\,\text{GeV}^4, \qquad \alpha_s \langle \bar{q}q \rangle^2 = (4 \pm 4) \cdot 10^{-4}\,\text{GeV}^6. \tag{14}$$

For the errors coming from the uncertainty of the QCD scale we take

$$\Lambda_{\overline{\text{MS}}} = 380 \pm 60\,\text{MeV} \tag{15}$$

The errors resulting from the uncertainty in the QCD scale in different energy intervals are clearly correlated and will have to be added linearly in the end. We also include the error of the strange quark mass in the light quark region which is taken as

$$\bar{m}_s(1\,\text{GeV}) = 200 \pm 60\,\text{MeV} \tag{16}$$

For the charm and bottom quark masses we use the values

$$\bar{m}_c(m_c) = 1.4 \pm 0.2\,\text{GeV}, \qquad \bar{m}_b(m_b) = 4.8 \pm 0.3\,\text{GeV}. \tag{17}$$

Summing up the contributions from the five flavours u, d, s, c, and b, our result for the hadronic contribution to the dispersion integral including the systematic error due to the dependence on $\Lambda_{\overline{\text{MS}}}$ (column 5 in Table 1) reads

$$\Delta\alpha_{\text{had}}^{(5)}(M_Z) = (277.6 \pm 4.1) \cdot 10^{-4}. \tag{18}$$

In order to obtain the total result for $\alpha(M_Z)$, we have to add the lepton and top contributions. Since we have nothing new to add to the calculation of these contributions we simply take the values from ref. [5],

$$\Delta\alpha_{\text{had}}^t(M_Z) = (-0.70 \pm 0.05) \cdot 10^{-4}, \qquad \Delta\alpha_{\text{lep}}(M_Z) \approx 314.97 \cdot 10^{-4}. \tag{19}$$

Writing $\Delta\alpha(M_Z) = \Delta\alpha_{\text{lep}}(M_Z) + \Delta\alpha_{\text{had}}(M_Z)$ our final result is ($\alpha(0)^{-1} = 137.036$)

$$\alpha(M_Z)^{-1} = \alpha(0)^{-1}(1 - \Delta\alpha(M_Z)) = 128.925 \pm 0.056. \tag{20}$$

CONCLUSION AND OUTLOOK

I have presented a method to obtain the running fine structure constant α at the scale of the Z_0 mass with minimal input of experimental data. This method is conservative in the meaning that its error is as free from assumptions as possible. Our method is discussed and compared with other methods (see e.g. ref. [13] – however, with our preliminary results). In ref. [14] Matthias Steinhauser says that "it is very impressive that the new analysis show very good agreement both in their central values and their quoted errors." I nevertheless would like to close this talk with the remark that all recent calculations of $\alpha(M_Z)$ should not deter experimentalists from remeasuring the e^+e^- annihilation cross section more accurately in the low and intermediate energy region, as such data are absolutely essential for a precise value of $\alpha(M_Z)$, unbiased by theory.

TABLE 1. Contributions of different energy intervals to $\alpha_{\text{had}}^{(5)}(M_Z)$. Second column: choice of neighbouring pairs of the polynomial degree N. Third column: fraction of the contribution of experimental data [1]. Fourth column: contribution to $\Delta\alpha_{\text{had}}^{(5)}(M_Z)$ with all errors included except for the systematic error due to the dependence on $\Lambda_{\overline{\text{MS}}}$ which is separately listed in the fifth column.

interval for $\sqrt{s}$	values of N	data contribution	contribution to $\Delta\alpha_{\text{had}}^{(5)}(M_Z)$	error due to $\Lambda_{\overline{\text{MS}}}$
$[0.28\,\text{GeV}, 3.1\,\text{GeV}]$	1, 2	24%	$(73.9 \pm 1.1) \cdot 10^{-4}$	$0.9 \cdot 10^{-4}$
$[3.1\,\text{GeV}, 9.46\,\text{GeV}]$	3, 4	0.3%	$(69.5 \pm 3.0) \cdot 10^{-4}$	$1.4 \cdot 10^{-4}$
$[9.46\,\text{GeV}, 30\,\text{GeV}]$	3, 4	1.1%	$(71.6 \pm 0.5) \cdot 10^{-4}$	$0.06 \cdot 10^{-4}$
$[30\,\text{GeV}, 40\,\text{GeV}]$	3, 4	0.15%	$(19.93 \pm 0.01) \cdot 10^{-4}$	$0.02 \cdot 10^{-4}$
$\sqrt{s} > 40\,\text{GeV}$			$(42.67 \pm 0.09) \cdot 10^{-4}$	
total range			$(277.6 \pm 3.2) \cdot 10^{-4}$	$1.67 \cdot 10^{-4}$

ACKNOWLEDGEMENTS

We would like to thank F. Jegerlehner and R. Harlander for providing us with material in addition to refs. [1,7], A.H. Hoang for correspondence, and G. Quast for discussions on all experimental aspects of this work and for continuing encouragement. S.G. gratefully acknowledges a grant given by the Max Kade Foundation.

REFERENCES

1. S. Eidelman and F. Jegerlehner, *Z. Phys.* **C67**, 585 (1995)
2. H. Burkhardt and B. Pietrzyk, *Phys. Lett.* **356 B**, 398 (1995)
3. M. Davier and A. Höcker, *Phys. Lett.* **419 B**, 419 (1998)
4. M.L. Swartz, *Phys. Rev.* **D53**, 5268 (1996)
5. J.H. Kühn and M. Steinhauser, *Phys. Lett.* **437 B**, 425 (1998)
6. S. Groote, J.G. Körner, K. Schilcher and N.F. Nasrallah, *Phys. Lett.* **440 B**, 375 (1998)
7. K.G. Chetyrkin, R. Harlander, J.H. Kühn and M. Steinhauser, *Nucl. Phys.* **B503**, 339 (1997)
8. S.G. Gorishny, A.L. Kataev and S.A. Larin, *Nuovo Cimento* **92 A**, 119 (1986)
9. K.G. Chetyrkin, B.A. Kniehl and M. Steinhauser, *Phys. Rev. Lett.* **79**, 353 (1997); *ibid.*, Report No. MPI/PhT/97-041, hep-ph/9708255
10. N. Gray, D.J. Broadhurst, W. Grafe and K. Schilcher, *Z. Phys.* **C48**, 673 (1990)
11. K.G. Chetyrkin, *Phys. Lett.* **404 B**, 161 (1997)
12. K.G. Chetyrkin, D. Pirjol and K. Schilcher, *Phys. Lett.* **404 B**, 337 (1997)
13. M. Davier and A. Höcker, *Phys. Lett.* **435 B**, 427 (1998)
14. M. Steinhauser, "Hadronic contribution to $\alpha(M_Z^2)$ and the anomalous magnetic moment of the muon", talk given at the 17th International Workshop on Weak Interactions and Neutrinos (WIN 99), Cape Town, South Africa, January 24–30, 1999, Report No. BUTP-99/06, hep-ph/9904373

Looking for New Physics in B_d^0-$\overline{B_d^0}$ Mixing[1]

David London

Laboratoire de René J.-A. Lévesque, Université de Montréal
C.P. 6128, succ. centre-ville, Montréal, QC, Canada

Abstract. There are variety of methods which directly test for the presence of new physics in the $b \to s$ flavour-changing neutral current (FCNC), but none which cleanly probe new physics in the $b \to d$ FCNC. One possible idea is to compare the weak phase of the t-quark contribution to the $b \to d$ penguin, which is $-\beta$ in the SM, with that of B_d^0-$\overline{B_d^0}$ mixing (-2β in the SM). In this talk I show that, in fact, it is impossible to measure the weak phase of the t-quark penguin, or indeed any penguin contribution, without theoretical input. However, if one makes a single assumption involving the hadronic parameters, it is possible to obtain the weak phase. I discuss how one can apply such an assumption to the time-dependent decays $B_d^0(t) \to K^0 \overline{K^0}$ and $B_s^0(t) \to \phi K_S$ in order to detect new physics in the $b \to d$ FCNC.

In the coming years, experiments at B-factories, HERA-B and hadron colliders will measure CP-violating asymmetries in B decays [1]. As always, the goal is to test the predictions of the standard model (SM). If we are lucky, there will be an inconsistency with the SM, thereby revealing the presence of new physics.

In the SM, CP violation is due to a complex phase in the Cabibbo-Kobayashi-Maskawa (CKM) mixing matrix. In the Wolfenstein parametrization of the CKM matrix [2], only the elements V_{ub} and V_{td} have non-negligible phases:

$$V_{CKM} = \begin{pmatrix} 1 - \frac{1}{2}\lambda^2 & \lambda & A\lambda^3(\rho - i\eta) \\ -\lambda & 1 - \frac{1}{2}\lambda^2 & A\lambda^2 \\ A\lambda^3(1 - \rho - i\eta) & -A\lambda^2 & 1 \end{pmatrix}. \quad (1)$$

It is convenient to parametrize V_{ub} and V_{td} as follows:

$$V_{ub} = |V_{ub}|e^{-i\gamma} , \quad V_{td} = |V_{td}|e^{-i\beta} . \quad (2)$$

Even though these elements are written in terms of two complex phases β and γ, it must be remembered that in fact there is only a single phase η in the CKM matrix; if η were to vanish, both β and γ would vanish as well.

---

[1] Talk based on work done in collaboration with A. Ali, N. Sinha and R. Sinha, and C.S. Kim and T. Yoshikawa.

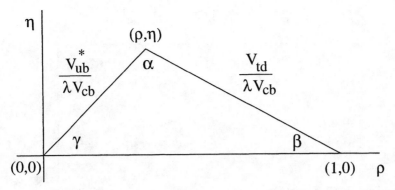

FIGURE 1. The unitarity triangle. The angles α, β and γ can be measured via CP violation in the B system.

The phase information in the CKM matrix can be elegantly displayed using the unitarity triangle [3]. The orthogonality of the first and third columns gives

$$V_{ud}V_{ub}^* + V_{cd}V_{cb}^* + V_{td}V_{tb}^* = 0 \ , \tag{3}$$

which is a triangle relation in the complex ρ-η plane, shown in Fig. 1. The angles β and γ are two of the interior angles of the unitarity triangle, with the third angle α satisfying $\alpha + \beta + \gamma = \pi$.

There are a variety of constraints on the unitarity triangle coming from (i) the extraction of $|V_{cb}|$ and $|V_{ub}|$ from semileptonic B decays, (ii) the measurements of $|V_{td}|$ and $|V_{ts}|$ in B_d^0-$\overline{B_d^0}$ and B_s^0-$\overline{B_s^0}$ mixing, and (iii) CP violation in the kaon system (ϵ). Unfortunately, there are substantial theoretical uncertainties in all of these constraints. For example, the theoretical expressions for ϵ and B_d^0-$\overline{B_d^0}$ mixing depend respectively on the bag parameter $B_K = 0.94 \pm 0.15$ and $f_{B_d}\sqrt{B_{B_d}} = 215 \pm 40$ MeV. The estimates of the magnitudes of these errors, which lie in the range 15–20%, come mainly from lattice calculations. Combining the experimental errors and theoretical uncertainties in quadrature [4], the presently-allowed region of the unitarity triangle is shown in Fig. 2. Due to the theoretical uncertainties, we do not have precise SM predictions for the CP phases α, β and γ: instead, these phases can take a range of values.

Since the hope is to find physics beyond the SM, the first question to be answered is: how can new physics affect the CP-violating asymmetries? There are two possible ways: the new physics can affect B decays or B mixing. Now, most B decays are dominated by a W-mediated tree-level diagram. In most models of new physics, there are no contributions to B decays which can compete with the SM. Thus, in general, the new physics cannot significantly affect the decays[2]. However,

---

[2] There is an exception: if the decay process is dominated by a penguin diagram, rather than a

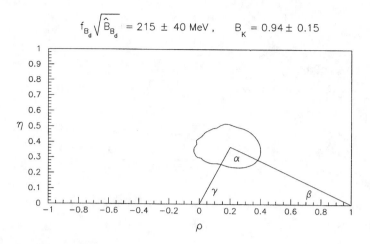

FIGURE 2. Allowed region (95% C.L.) in the ρ-η plane, from a simultaneous fit to all experimental and theoretical data. The theoretical errors are treated as Gaussian for this fit. The triangle shows the best fit.

the CP asymmetries *can* be affected if there are new contributions to B^0-$\overline{B^0}$ mixing [6]. The presence of such new-physics contributions will affect the extraction of V_{td} and V_{ts}. And if there are new phases, the measurements of α, β and γ will also be affected. Thus, new physics enters principally through new contributions to B^0-$\overline{B^0}$ mixing [7].

Unfortunately, this creates a bit of a problem. B-factories such as BaBar and Belle will measure α, β and γ via $B_d^0(t) \to \pi^+\pi^-$ (or $\rho\pi$ [8]), $B_d^0(t) \to \Psi K_s$, and $B^\pm \to DK^\pm$ [9], respectively. Note that only the first two decays involve B^0-$\overline{B^0}$ mixing. Thus, if there is new physics, only the measurements of α and β will be affected. However, they will be affected in opposite directions [10]. That is, in the presence of a new-physics phase ϕ_{NP}, the CP angles are changed as follows: $\alpha \to \alpha + \phi_{NP}$ and $\beta \to \beta - \phi_{NP}$. The key point is that ϕ_{NP} cancels in the sum $\alpha + \beta + \gamma$, so that this sum is *insensitive* to the new physics, i.e. B-factories will always find $\alpha + \beta + \gamma = \pi$. (Note that hadron colliders do not suffer from the same problem – if γ is measured in $B_s^0(t) \to D_s^\pm K^\mp$ [11], then $\alpha + \beta + \gamma \neq \pi$ can be found if there is new physics in B_s^0-$\overline{B_s^0}$ mixing.)

Thus, B-factories cannot discover new physics via $\alpha + \beta + \gamma \neq \pi$. Still, new physics can be found if the measurements of the angles are inconsistent with the measurements of the sides. However:

1. the allowed region of the unitarity triangle is still fairly large. It is conceivable that even in the presence of new physics, the triangle as constructed from the angles α, β and γ will still lie within the allowed region;

---

tree-level diagram, then new physics *can* significantly affect the decay, see Refs. [1,5].

2. even if the α-β-γ triangle lies outside the allowed region, is this evidence of new physics, or have we underestimated the theoretical uncertainties which go into the constraints of the unitarity triangle (Fig. 2)?

The point is: ideally, we would like cleaner, more direct tests of the SM in order to probe for the presence of new physics.

In fact, there are such direct tests:

1. $B^\pm \to DK^\pm$ vs. $B_s^0(t) \to D_s^\pm K^\mp$: in the SM, both of these CP asymmetries measure γ. If there is a discrepancy between the value of γ as extracted from these two decays, this points to new physics in B_s^0-$\overline{B_s^0}$ mixing.

2. $B_d^0(t) \to \Psi K_S$ vs. $B_d^0(t) \to \phi K_S$: in the SM, both of these decays measure β. A discrepancy implies new physics in the $b \to s$ penguin [5].

3. $B_s^0(t) \to \Psi\phi$: in the SM, the CP asymmetry in this decay vanishes (to a good approximation). If this CP asymmetry is found to be nonzero, this again indicates the presence of new physics in B_s^0-$\overline{B_s^0}$ mixing.

There are thus several direct tests for new physics. However, note: all of these tests probe new physics in either B_s^0-$\overline{B_s^0}$ mixing or the $b \to s$ penguin, i.e. in the $b \to s$ flavour-changing neutral current (FCNC).

So this raises the question: are there any direct tests of new physics in the $b \to d$ FCNC?

Consider pure $b \to d$ penguin decays such as $B_d^0 \to K^0 \overline{K^0}$ or $B_s^0 \to \phi K_S$. Such decays involve up-type quarks in the loop. If the t-quark contribution dominated, then the $b \to d$ penguin amplitude would be proportional to $V_{tb}^* V_{td}$. Recalling that the weak phase of B_d^0-$\overline{B_d^0}$ mixing is -2β and that the weak phase of V_{td} is $-\beta$, in such a case the SM would predict that (i) the CP asymmetry in $B_d^0(t) \to K^0 \overline{K^0}$ vanishes, and (ii) the CP asymmetry in $B_s^0(t) \to \phi K_S$ measures $\sin 2\beta$ [12]. Any discrepancy between measurements of these CP asymmetries and their predictions would thus imply that there is new physics in either B_d^0-$\overline{B_d^0}$ mixing or the $b \to d$ penguin, i.e. in the $b \to d$ FCNC. (In the second decay, new physics in B_s^0-$\overline{B_s^0}$ mixing could also come into play, but that can be established independently, as discussed above).

However, $b \to d$ penguins are *not* dominated by the internal t-quark. The contributions of the u- and c-quarks can be as large as 20–50% of that of the t-quark [13]. In this case, the above predictions of the SM no longer hold, so that one cannot test for new physics in the $b \to d$ FCNC in this way.

So this raises a new question: are there ways of cleanly measuring the weak phase of the t-quark contribution to the $b \to d$ penguin? Unfortunately, the answer to this question is *no* [14].

To see this, consider the general form of the amplitude for the $b \to d$ penguin. There are three terms, corresponding to the contributions of the three internal up-type quarks:

$$P = P_u V_{ub}^* V_{ud} + P_c V_{cb}^* V_{cd} + P_t V_{tb}^* V_{td}, \qquad (4)$$

and recall that $V_{ub} \sim e^{-i\gamma}$ and $V_{td} \sim e^{-i\beta}$.

Using the unitarity relation of Eq. 3, the u-quark piece can be eliminated in the above equation, allowing us to write

$$P = \mathcal{P}_{cu} e^{i\delta_{cu}} + \mathcal{P}_{tu} e^{i\delta_{tu}} e^{-i\beta} , \qquad (5)$$

where δ_{cu} and δ_{tu} are strong phases. Now imagine that there were a method in which a series of measurements allowed us to cleanly extract β using the above expression. In this case, we would be able to express $-\beta$ as a function of the observables.

On the other hand, we can instead use the unitarity relation to eliminate the t-quark contribution in Eq. 4, yielding

$$P = \mathcal{P}_{ct} e^{i\delta_{ct}} + \mathcal{P}_{ut} e^{i\delta_{ut}} e^{i\gamma} . \qquad (6)$$

Comparing Eqs. 5 and 6, we see that they have the same form. Thus, the same method which allowed us to extract $-\beta$ from Eq. 5 should be applicable to Eq. 6, allowing us to obtain γ. That is, we would be able to write γ as *the same function of the observables* as was used for $-\beta$ above! But this implies that $-\beta = \gamma$, which clearly doesn't hold in general.

Due to the ambiguity in the parametrization of the $b \to d$ penguin — which I will refer to as the *CKM ambiguity* — we therefore conclude that one cannot cleanly extract the weak phase of any penguin contribution. Indeed, it is *impossible* to cleanly test for the presence of new physics in the $b \to d$ FCNC.

Nevertheless, it is interesting to examine some candidate methods and see how they fail. For example, consider the time-dependent rate for the decay $B_d^0(t) \to K^0 \overline{K^0}$. This can be written

$$\Gamma(B_d^0(t) \to K^0 \overline{K^0}) = e^{-\Gamma t} \left[\frac{|A|^2 + |\bar{A}|^2}{2} + \frac{|A|^2 - |\bar{A}|^2}{2} \cos(\Delta M t) \right.$$
$$\left. - \mathrm{Im}\left(e^{-2i\beta} A^* \bar{A}\right) \sin(\Delta M t) \right], \qquad (7)$$

where $A \equiv A(B_d^0 \to K^0 \overline{K^0})$ and $\bar{A} \equiv A(\overline{B_d^0} \to K^0 \overline{K^0})$. The measurement of this time-dependent decay rate allows one to extract the magnitudes and relative phase of $e^{i\beta} A$ and $e^{-i\beta} \bar{A}$. Using the form of the $b \to d$ penguin given in Eq. 5, we have

$$e^{i\beta} A = e^{i\beta} \left[\mathcal{P}_{cu} e^{i\delta_{cu}} + \mathcal{P}_{tu} e^{i\delta_{tu}} e^{-i\beta'} \right], \qquad (8)$$

where in the last term I have written the weak phase as β' to allow for the possibility of new physics in the $b \to d$ FCNC. There are thus 5 measurable parameters: $\mathcal{P}_{cu}$, $\mathcal{P}_{tu}$, $\delta_{cu} - \delta_{tu}$, β, and $\theta_{NP} \equiv \beta' - \beta$. However, there are only 4 measurements: the coefficients of the 3 time-dependent functions $[1, \cos(\Delta M t), \sin(\Delta M t)]$ in Eq. 7, and one independent measurement of β. Therefore, as argued above, there are not enough measurements to determine all the theoretical parameters. More to the point, there is one more theoretical unknown than there are measurements.

In fact, one can examine a variety of other techniques: $B \to \pi\pi$ isospin analysis [15], Dalitz plot analysis of $B \to 3\pi$ [8], angular analysis of the decay of a neutral B-meson to two vector mesons [16], and a combined isospin + angular analysis of $B \to \rho\rho$. In all cases there is one more unknown than there are measurements. From this we can therefore conclude the following: due to the CKM ambiguity, if we wish to test for the presence of new physics in the $b \to d$ FCNC by comparing the weak phase of B_d^0-$\overline{B_d^0}$ mixing with that of the t-quark contribution to the $b \to d$ penguin, it is necessary to make a single assumption about the theoretical (hadronic) parameters describing the decay.

As an example of such an assumption, consider again the two decays $B_d^0(t) \to K^0\overline{K^0}$ and $B_s^0(t) \to \phi K_S$. Recall that we can write the $B_d^0 \to K^0\overline{K^0}$ amplitude as

$$e^{i\beta} A_d^{K^0\overline{K^0}} = \mathcal{P}_{cu} e^{i\delta_{cu}} e^{i\beta} + \mathcal{P}_{tu} e^{i\delta_{tu}} e^{-i(\beta'-\beta)} . \tag{9}$$

Assuming that there is no new physics in B_s^0-$\overline{B_s^0}$ mixing, we can write the $B_s^0 \to \phi K_S$ amplitude as

$$A_s^{\phi K_S} = \tilde{\mathcal{P}}_{cu} e^{i\tilde{\delta}_{cu}} + \tilde{\mathcal{P}}_{tu} e^{i\tilde{\delta}_{tu}} e^{-i\beta'} . \tag{10}$$

The tildes are added to distinguish the parameters in the decay $B_s^0 \to \phi K_S$ from those in $B_d^0 \to K^0\overline{K^0}$. There are two reasons. First, in the B_s^0 decay, we have a spectator s-quark instead of a d-quark. And second, there are colour-allowed electroweak penguin contributions to $B_s^0 \to \phi K_S$ while there are none in $B_d^0 \to K^0\overline{K^0}$.

From the above, we see that there are 8 theoretical parameters describing these two decays: $\mathcal{P}_{cu}$, $\mathcal{P}_{tu}$, $\tilde{\mathcal{P}}_{cu}$, $\tilde{\mathcal{P}}_{tu}$, β, β', $\delta_{cu} - \delta_{tu}$, and $\tilde{\delta}_{cu} - \tilde{\delta}_{tu}$. However there are only 7 experimental measurements: the magnitudes and relative phase of $e^{i\beta} A_d^{K^0\overline{K^0}}$ and $e^{-i\beta} \bar{A}_d^{K^0\overline{K^0}}$, the magnitudes and relative phase of $A_s^{\phi K_S}$ and $\bar{A}_s^{\phi K_S}$, and an independent measurement of β. If we wish to determine the theoretical parameters, we therefore need to make an assumption.

In Ref. [17], the following assumption is made. Defining $r \equiv \mathcal{P}_{cu}/\mathcal{P}_{tu}$ and $\tilde{r} \equiv \tilde{\mathcal{P}}_{cu}/\tilde{\mathcal{P}}_{tu}$, it is assumed that $r = \tilde{r}$. How good is this assumption? Writing

$$r = \left| \frac{P_c - P_u}{P_t - P_u + P_{EW}^C} \right| , \quad \tilde{r} = \left| \frac{\tilde{P}_c - \tilde{P}_u}{\tilde{P}_t - \tilde{P}_u + \tilde{P}_{EW} + \tilde{P}_{EW}^C} \right| , \tag{11}$$

we note the following. Since the spectator-quark effects cancel in the ratio in $\tilde{r}$, the principle difference between r and $\tilde{r}$ is due to the presence of the colour-allowed electroweak penguin contribution in the denominator of $\tilde{r}$. Since $\tilde{P}_{EW}/\tilde{P}_t \simeq 20\%$, we therefore conclude that r and $\tilde{r}$ are equal to within roughly 20%. Taking $r = \tilde{r}$ is therefore a reasonable assumption.

With this assumption, we now have an equal number of theoretical unknowns and experimental measurements, and can therefore solve for β and β' independently. In this way we can test for the presence of new physics in the $b \to d$ FCNC. Note also that the assumption of $r = \tilde{r}$ holds only within a particular parametrization of the $b \to d$ penguin, so that the CKM ambiguity is lifted.

There are, in fact, other methods where an assumption can be used to measure the weak phase of the t-quark contribution to the $b \to d$ penguin. My collaborators and I are currently examining such methods.

To summarize: if the unitarity triangle as constructed from measurements of the CP angles α, β and γ disagrees with that constructed from measurements of the sides, we may deduce that there is new physics in B_d^0-$\overline{B_d^0}$ mixing. However, it may be that the discrepancy is due not to the presence of new physics, but rather to an underestimate of the theoretical uncertainties which enter into the constraints on the unitarity triangle. For this reason, it is preferable to have direct tests for new physics.

There are, in fact, several such direct tests, but they all probe new physics in the $b \to s$ FCNC. One possibility of searching for new physics in the $b \to d$ FCNC is the following: in the SM the weak phase of B_d^0-$\overline{B_d^0}$ mixing is -2β, while that of the t-quark contribution to the $b \to d$ penguin is $-\beta$. A comparison of these two weak phases might reveal new physics in the $b \to d$ FCNC.

Unfortunately, due to the ambiguity in parametrizing the $b \to d$ penguin, it is impossible to cleanly measure the weak phase of the t-quark contribution to the $b \to d$ penguin. In order to measure this phase, it is necessary to make an assumption about the hadronic parameters. I presented one example involving the two decays $B_d^0(t) \to K^0 \overline{K^0}$ and $B_s^0(t) \to \phi K_S$, but there are other methods. With such an assumption it is possible to detect the presence of new physics in the $b \to d$ FCNC.

Acknowledgments

I would like to thank the organizers of MRST '99 for a very enjoyable conference. This research was financially supported by NSERC of Canada and FCAR du Québec.

REFERENCES

1. For a review of CP violation in the B system, see, for example, *The BaBar Physics Book*, eds. P.F. Harrison and H.R. Quinn, SLAC Report 504, October 1998.
2. L. Wolfenstein, *Phys. Rev. Lett.* **51**, 1945 (1983).
3. C. Caso et al. (Particle Data Group), *Eur. Phys. J.* **C3**, 1 (1998).
4. A. Ali and D. London, hep-ph/9903535, to be published in the *Eur. Phys. J.* **C**, 1999.
5. Y. Grossman and M.P. Worah, *Phys. Lett.* **395B**, 241 (1997); D. London and A. Soni, *Phys. Lett.* **407B**, 61 (1997).
6. C.O. Dib, D. London and Y. Nir, *Int. J. Mod. Phys.* **A6**, 1253 (1991).
7. For a review of new-physics effects in CP asymmetries in the B system, see M. Gronau and D. London, *Phys. Rev.* **D55**, 2845 (1997), and references therein.

8. A.E. Snyder and H.R. Quinn, *Phys. Rev.* **D48**, 2139 (1993).
9. M. Gronau and D. Wyler, *Phys. Lett.* **265B**, 172 (1991). See also M. Gronau and D. London, *Phys. Lett.* **253B**, 483 (1991); I. Dunietz, *Phys. Lett.* **270B**, 75 (1991). Improvements to this method have recently been discussed by D. Atwood, I. Dunietz and A. Soni, *Phys. Rev. Lett.* **78**, 3257 (1997).
10. Y. Nir and D. Silverman, *Nucl. Phys.* **B345**, 301 (1990).
11. R. Aleksan, I. Dunietz, B. Kayser and F. Le Diberder, *Nucl. Phys.* **B361**, 141 (1991); R. Aleksan, I. Dunietz and B. Kayser, *Zeit. Phys.* **C54**, 653 (1992).
12. D. London and R. Peccei, *Phys. Lett.* **223B**, 257 (1989).
13. A.J. Buras and R. Fleischer, *Phys. Lett.* **341B**, 379 (1995).
14. D. London, N. Sinha and R. Sinha, hep-ph/9905404, to be published in *Phys. Rev. D*, 1999.
15. M. Gronau and D. London, *Phys. Rev. Lett.* **65**, 3381 (1990).
16. I. Dunietz, H.R. Quinn, A. Snyder, W. Toki and H.J. Lipkin, *Phys. Rev.* **D43**, 2193 (1991).
17. C.S. Kim, D. London and T. Yoshikawa, hep-ph/9904311, to be published in *Phys. Lett. B*, 1999.

T-Odd Triple Product Asymmetries in Beauty Decays

Wafia Bensalem

Laboratoire René J.-A. Lévesque, Université de Montréal
C.P. 6128, succ. centre-ville, Montréal, QC, Canada H3C 3J7

Abstract. The decay $b \to su\bar{u}$ receives contributions from both tree and penguin amplitudes. In this talk I show that tree-penguin interference gives rise to a T-violating triple product $\vec{s}_b \cdot (\vec{p}_u \times \vec{p}_s)$ between the b-quark spin and the momenta of the s and u quarks. In the standard model, the T-violating signal turns out to be rather large, about 6% of the total rate. This is quite encouraging, and suggests that triple products may be useful for testing the standard model and searching for new physics.

INTRODUCTION

Since the discovery of CP violation in the K system, there remains a burning question: is the standard model (SM) explanation of CP violation (a complex phase in the Cabibbo-Kobayashi-Maskawa (CKM) matrix) the correct one? To answer this question, we should look for as many CP-violating signals as possible, both within and outside the K system, in order to better understand and to carefully test the SM. Beauty decays offer many possibilities for detecting CP violation. The most well-known of these are rate asymmetries [1]. However, there is another possibility: signals of triple products.

Triple products (TP's) are T-odd correlations of three vectors of the form:

$$TP = \vec{v}_1 \cdot (\vec{v}_2 \times \vec{v}_3), \qquad (1)$$

where each vector is the spin or momentum of a particle participating in the decay process. Assuming that the CPT theorem is valid, T-odd observables are equivalent to CP-odd observables, and hence the presence of any triple product in the partial decay rate is a signal of both CP and T violation [2].

The CLEO collaboration has measured large branching ratios for processes involving $b \to sq\bar{q}$ inclusive modes [3]. Among these, the mode $b \to su\bar{u}$ receives both tree and penguin contributions. This makes this decay a good potential candidate to exhibit T-odd triple products due to the interference between the two diagrams. Indeed, as I will show, there *is* a triple product in the partial decay rate of this mode, corresponding to an asymmetry of about 6% in the SM.

CP/T VIOLATION VIA TRIPLE PRODUCTS

We know that time reversal symmetry, T, reverses the signs of both spin and momentum vectors. Hence, triple products of the form given in Eq. (1) are reversed under T. Thus, if the partial decay rate of a process contains a TP, the process is T/CP violating.

T-Odd Asymmetry

Once we find a triple product in the partial decay rate, we will want to estimate the size of the T-violating effect. To do so, we define an asymmetry by dividing the phase space into two regions depending on the sign of the triple product [4]. The T-odd asymmetry[1] is defined by:

$$A_T \equiv \frac{\Gamma(TP > 0) - \Gamma(TP < 0)}{\Gamma(TP > 0) + \Gamma(TP < 0)}, \tag{2}$$

where Γ is the total decay rate.

Distinguishing the true signal of T violation

As I will illustrate in the next section, the T-odd asymmetry has the general form:

$$A_T \propto \sin\phi_W \cos\delta_S + \cos\phi_W \sin\delta_S, \tag{3}$$

where δ_S is the strong phase and ϕ_W is the weak phase which is responsible for CP violation. Thus, if $\phi_W = 0$ but $\delta_S \neq 0$, one would have a nonzero A_T, despite the fact that there is no CP violation. That is, the presence of strong phases allows for the possibility of a fake signal of T violation.

In order to distinguish the true signal of T/CP violation from the fake T-odd effect due only to strong phases, we note that when one compares a decay $b \to f$ with its conjugate $\bar{b} \to \bar{f}$, the weak phases change sign, but the strong phases do not. We can therefore consider the true **T-violating asymmetry** defined by:

$$\mathcal{A}_T = \frac{1}{2}(A_T - \bar{A}_T), \tag{4}$$

where $\bar{A}_T$ is the CP-conjugate T-odd asymmetry. Thus,

$$\mathcal{A}_T \propto \sin\phi_W \cos\delta_S. \tag{5}$$

This is a true T/CP-violating signal in that $\mathcal{A}_T$ differs from zero only if $\phi_W \neq 0$, i.e. if T/CP is violated in the considered decay.

---

[1] If a triple product changes sign, it is not necessarily signal of T violation. This is because, in addition to reversing spins and momenta, the time reversal symmetry T also exchanges the initial and final states. For this reason I refer to the asymmetry as a T-odd effect, rather than a T-violating one. I show below how to establish the presence of a true signal of T violation.

TRIPLE PRODUCTS IN $b \to su\bar{u}$

In the inclusive decay $b \to su\bar{u}$, the amplitude has two dominant contributions: the tree diagram (T) due to W-boson exchange and the loop-level strong penguin diagram (P). The first step in the calculation of the decay rate is the loop calculation of the penguin amplitude, which contains two dominant terms:

$$P \simeq P_1 + P_2 , \tag{6}$$

with

$$P_1 = \frac{i\alpha_s G_F}{\sqrt{2}\pi} F_1^c V_{cs}^* V_{cb} \left[\bar{s}t^\alpha \gamma_\mu \gamma_L b\right]\left[\bar{u}t_\alpha \gamma^\mu v_u\right] e^{i\delta_1} , \tag{7}$$

$$P_2 = \frac{i\alpha_s G_F}{\sqrt{2}\pi} \left[\frac{-im_b}{q^2} F_2\right] V_{ts}^* V_{tb} \left[\bar{s}t^\alpha \sigma_{\mu\nu} q^\nu \gamma_R b\right]\left[\bar{u}t_\alpha \gamma^\mu v_u\right] e^{i\delta_2} . \tag{8}$$

In the above, α_s and G_F are the strong and Fermi coupling constants, respectively, V_{ij} are the CKM matrix elements, the t^α are the Gell-Mann matrices, and the δ_i are the strong phases. q is the momentum of the internal gluon. F_1^c and F_2 are functions of (m_c^2/M_W^2) and (m_t^2/M_W^2), respectively, and take the values $F_1^c \simeq 5.0$ and $F_2 \simeq 0.2$ for $m_t = 160$ GeV [5]. P_2 is often called *the chromomagnetic dipole moment* term and, as we shall see, is responsible for T/CP violation.

The next step is the calculation of the square of the decay amplitude, $|\mathcal{M}|^2$. We have:

$$\begin{aligned}|\mathcal{M}|^2 &= |T + P_1 + P_2|^2 \\ &= |T|^2 + |P_1|^2 + |P_2|^2 + 2Re\left(T^\dagger P_1\right) + 2Re\left(T^\dagger P_2\right) + 2Re\left(P_1^\dagger P_2\right) . \end{aligned} \tag{9}$$

We find triple products in the both of the interference terms involving the chromomagnetic dipole operator (i.e. the two last terms of $|\mathcal{M}|^2$).

Fixing the polarization of the b quark, and summing over the spins of the s, u and $\bar{u}$ quarks[2], we find

$$\sum_{u,\bar{u},s\ spins} 2Re\left(P_1^\dagger P_2\right) = \frac{16\alpha_s^2 G_F^2 F_1^c F_2 m_b}{3\pi^2 q^2} Im\left[V_{ts}^* V_{tb} V_{cs} V_{cb}^* e^{i(\delta_2 - \delta_1)}\right]$$
$$\times p_b \cdot (p_u - p_{\bar{u}})\ \epsilon_{\mu\nu\rho\xi}\ p_b^\mu s_b^\nu p_u^\rho p_s^\xi + ... , \tag{10}$$

$$\sum_{u,\bar{u},s\ spins} 2Re\left(T^\dagger P_2\right) = \frac{128\alpha_s G_F^2 F_2 m_b}{3\pi q^2} Im\left[V_{ts}^* V_{tb} V_{us} V_{ub}^* e^{i(\delta_2 - \delta_t)}\right]$$
$$\times p_s \cdot p_u\ \epsilon_{\mu\nu\rho\xi}\ p_b^\mu s_b^\nu p_u^\rho p_s^\xi + ... \tag{11}$$

---

[2] Triple products involving the polarization of the s quark are suppressed by at least m_s/m_b (and similarly for the u and $\bar{u}$ quarks), and so are negligible.

In the above, the ellipsis (...) indicates terms which are negligible compared to $|P_1|^2$. The triple product arises from the term $\epsilon_{\mu\nu\rho\xi} p_b^\mu s_b^\nu p_u^\rho p_s^\xi$, which, in the rest frame of the b quark, is exactly $m_b \vec{s}_b \cdot (\vec{p}_u \times \vec{p}_s)$.

Integrating over the phase space, we find that the penguin-penguin triple product is negligible compared to that due to tree-penguin interference. We can write the tree-penguin signal as follows:

$$\left[\sum_{u,\bar{u},s\ spins} 2 Re\left(T^\dagger P_2\right) \right]_{eff} = \frac{128\alpha_s G_F^2 F_2 m_b^2}{3\pi q^2} A^2 \lambda^6 \sigma p_s \cdot p_u \left[\vec{s}_b \cdot (\vec{p}_u \times \vec{p}_s) \right]$$
$$[\sin\delta \cos(\delta_2 - \delta_t) + \cos\delta \sin(\delta_2 - \delta_t)] \ . \quad (12)$$

A, λ, σ and δ are the CKM matrix parameters in the Wolfenstein representation. (σ and δ are related to the more commonly-used parameters ρ and η via $\sigma = \sqrt{\rho^2 + \eta^2}$ and $\tan\delta = \eta/\rho$.) Within the SM, CP violation is due solely to the presence of the complex phase δ (or, equivalently, η) in the CKM matrix.

In Eq. (12), we explicitly see the triple product involving the b-quark spin and the momenta of the s and u quarks. Note also that the second term in square brackets [in the second line of Eq. (12)] represents the fake signal of T/CP violation due to the strong phases. That is, we still have a T-odd effect even if the CP violating phase δ is zero.

But there is another important feature. In order to have CP violation in decay rate asymmetries $[a_{CP} = (\Gamma - \overline{\Gamma})/(\Gamma + \overline{\Gamma})]$ it is necessary that the strong phases be non-zero. In triple-product asymmetries, the presence of a strong phase is not necessary: we still have a signal when $(\delta_2 - \delta_t)$ is zero. In fact, the signal is maximal in this case.

Finally, to estimate the size of T/CP violation in $b \to su\bar{u}$, we must calculate the T-violating asymmetry for this process, defined in Eqs. (2) and (4). It is often assumed that strong phases in $b \to su\bar{u}$ come from the absorptive parts of the penguin contributions [6]. Since the triple product involves only the t-quark penguin contribution, which is purely dispersive, this leads to $(\delta_2 - \delta_t) = 0$. In the rest frame of the b quark, we then find

$$\mathcal{A}_T(b \to su\bar{u}) \simeq 6\% \ . \quad (13)$$

In the above, we have taken $\eta = 0.4$. (Using the presently-allowed range $0.26 \leq \eta \leq 0.52$ [7], we have $4\% \lesssim \mathcal{A}_T(b \to su\bar{u}) \lesssim 8\%$). This result can be compared with the decay rate asymmetry, calculated by Hou for the same process [8]:

$$a_{CP}(b \to su\bar{u}) \simeq 1.4\% \quad (14)$$

We therefore see that one expects the triple-product asymmetry in $b \to su\bar{u}$ to be considerably larger than the decay rate asymmetry.

APPLICATION

Given that the SM T-violating triple product asymmetry in $b \to su\bar{u}$ is large, the next obvious question is: in what exclusive beauty decays can one test this result?

It is clear that we cannot use decays of B mesons, since their spin is zero, and so there is no way to measure the spin of the b quark (which is the only spin contributing to the TP). However, one possibility would be to use the Λ_b baryon, whose spin is largely that of the b quark. For example, we can consider the process $\Lambda_b \to \Lambda_s \pi^0$. The TP found in the previous section can be roughly equated to $\vec{s}_{\Lambda_b} \cdot (\vec{p}_{\pi^0} \times \vec{p}_{\Lambda_s})$.

I note again that triple products involving the spins of the final particles are suppressed by powers of their masses. Hence, if a T-violating asymmetry due to a TP containing the spin of a final particle were measured in any beauty decay corresponding to the inclusive mode $b \to su\bar{u}$, that would probably indicate the presence of new physics. For example, one could examine the decay $B^+ \to \overline{\Lambda}_s p$ for triple products involving the spin of the $\bar{s}$ quark, such as $\vec{s}_{\bar{s}} \cdot (\vec{p}_u \times \vec{p}_{\bar{s}})$, which can be roughly equated with $\vec{s}_{\overline{\Lambda}_s} \cdot (\vec{p}_p \times \vec{p}_{\overline{\Lambda}_s})$.

CONCLUSION

In the standard model, we have found a significant signal (6%) of T/CP violation in the decay process $b \to su\bar{u}$. This signal is due to the presence of a triple product involving the spin of the b quark and the momenta of the s and the u quarks.

One nice thing about triple-product asymmetries is that, unlike direct CP-violating decay rate asymmetries, they are not dependent on non-zero strong phases. In fact, the asymmetries are maximal when the strong phases vanish.

T-violating triple product asymmetries may be useful for searching for CP violation beyond the SM. For example, in some models of new physics, the chromomagnetic dipole moment F_2 can be enhanced up to ten times its value in the SM. This will have an enormous effect on the triple product asymmetry. Also, triple-product signals involving the spins of the final-state particles (s, u and $\bar{u}$) are tiny in the SM, so that any measure of such signals will indicate new physics.

Acknowledgments

This research was financially supported by NSERC of Canada.

REFERENCES

1. For a review of CP violation in the B system, see, for example, *The BaBar Physics Book*, eds. P.F. Harrison and H.R. Quinn, SLAC Report 504, October 1998.
2. B. Kayser, *Nucl. Phys. Proc. Suppl.* **13**, 487 (1990).
3. T.E. Browder et al. (CLEO collaboration) *Phys. Rev. Lett.* **81**, 1786 (1998); Y. Gao and F. Würthwein, hep-ex/9904008.

4. G. Valencia, *Phys. Rev.* **D39**, 3339 (1989); E. Golowich, and G. Valencia, *Phys. Rev.* **D40**, 112 (1989).
5. W.-S. Hou, *Nucl. Phys.* **B308**, 561 (1988).
6. M. Bander, D. Silverman and A. Soni, *Phys. Rev. Lett.* **43**, 242 (1979).
7. A. Ali and D. London, hep-ph/9903535, to be published in the *Eur. Phys. J.* **C**, 1999.
8. W.-S. Hou, hep-ph/9902382.

Study of $B \to D^{(*)+}D^{(*)-}K_s$ Decays and the extraction of β

T. E. Browder[1] [b], Alakabha Datta[2] [a], Patrick. J. O'Donnell [3] [a] and Sandip Pakvasa [4] [b]

[a] *Department of Physics,*
University of Toronto, Toronto, Canada.
[b] *Department of Physics and Astronomy,*
University of Hawaii,
Honolulul, Hawaii
USA

Abstract. We consider the possibility of measuring $\sin(2\beta)$ in the KM unitarity triangle using the process $B^0 \to D^{*+}D^{*-}K_s$. This decay mode has a higher branching fraction (O(1%)) than the mode $B^0 \to D^{*+}D^{*-}$. We use the factorization assumption and heavy hadron chiral perturbation theory to estimate the branching fraction and polarization. The time dependent rate $B^0(t) \to D^{*+}D^{*-}K_s$ can be used to measure $\sin(2\beta)$ and $\cos(2\beta)$. Furthermore, examination of the $D^{*+}K_s$ mass spectrum may be the best way to experimentally find the broad 1^+ p-wave D_s meson.

INTRODUCTION

The decay $B^0 \to J/\psi K_s$ is expected to provide a clean measurement of the angle $\sin(2\beta)$ in the unitarity triangle [1]. However, other modes can also provide relevant information on the angle β. An example of such a mode is the decay $B^0 \to D^{(*)}\overline{D}^{(*)}$. The scalar final state in $B^0 \to D^+D^-$ is a definite CP eigenstate. For $B^0 \to D^{*+}D^{*-}$ the vector-vector final state can be CP even or CP odd state or an admixture. This is due to the fact that s, p and d partial waves with different CP-parities can contribute. An angular analysis can extract the contribution of the different CP eigenstates leading to a measurement of $\sin(2\beta)$ [2,3]. In the factorization assumption and using Heavy Quark Effective Theory (HQET) it can be shown that the final states in $B^0 \to D^{*+}D^{*-}$ is dominated by a single CP eigenstate [4]. If this is true, the angle $\sin(2\beta)$ can be determined without an angular analysis. The decay $B^0 \to D^{*+}D^{*-}$ may be preferred to $B^0 \to D^+D^-$ because

---

[1] email: teb@uhheph.phys.hawaii.edu
[2] email: datta@medb.physics.utoronto.ca
[3] email: pat@medb.physics.utoronto.ca
[4] email: pakvasa@uhheph.phys.hawaii.edu.

contamination from penguin contributions and final state interactions (FSI) are expected to be smaller in the former decay [3].

In this work we consider the possibility of extracting β from the decay $B^0 \to D^{(*)}\bar{D}^{(*)}K_s$. These decays are governed by the quark subprocess $b \to c\bar{c}s$ and so are enhanced relative to $B^0 \to D^{(*)}\overline{D}^{(*)}$ which are governed by the quark subprocess $b \to c\bar{c}d$ by the factor $|V_{cs}/V_{cd}|^2 \sim 20$. As in the case of $B^0 \to J/\psi K_s$ decay the penguin contamination is expected to be small in these decays.

Recently CLEO [5] has obtained $\mathcal{B}(\overline{B}^0 \to D^{*+}\bar{D}^{*0}K^-) = (1.30^{+0.61}_{-0.47} \pm 0.27)\%$ and $\mathcal{B}(B^- \to D^{*0}\bar{D}^{*0}K^-) = (1.45^{+0.78}_{-0.58} \pm 0.36)\%$. These values should be approximately equal to the branching fraction for $\mathcal{B}(B^0 \to D^{*+}D^{*-}K^0)$. We use the latter value for the purpose of a sensitivity estimate. Taking into account $\mathcal{B}(K^0 \to K_s) = 0.5$, $\mathcal{B}(K_s \to \pi^+\pi^-) = 0.667$, and assuming that the K_s reconstruction efficiency is ~ 0.5, we can estimate the ratio of the tagged $B^0 \to D^{*+}D^{*-}K_s$ events to the tagged $D^{*+}D^{*-}$ events. Assuming $\mathcal{B}(B^0 \to D^{*+}D^{*-}) = 6 \times 10^{-4}$, which is the central value of the recent CLEO measurement [6], we find that the ratio of the number of events is ~ 4.0. Therefore, this mode will be more sensitive to the CP violation angle $\sin(2\beta)$ than $B^0 \to D^{*+}D^{*-}$. A similar conclusion is obtained in the comparison of $B^0 \to D^+D^-K_s$ to $B^0 \to D^+D^-$. The above conclusions are detector dependent; a somewhat pessimistic estimate of the K_s reconstruction efficiency is used here while the detection efficiency for the $D^{*+}D^{*-}$ final state was assumed to be similar for both cases. Better determination of the CP sensitivities will require more precise measurements of the branching fractions for the $D^*\bar{D}^*K$ decay modes and will also depend on details of the experimental apparatus and reconstruction programs.

The amplitude for the decay $B^0 \to D^*\bar{D}^*K_s$ can have a resonant contribution and a non-resonant contribution. For the resonant contribution the D^*K_s in the final state comes dominantly from an excited $D_s(1^+)$ state. In the approximation of treating $D^*\bar{D}^*K_s$ as $D^*D_s(excited)$, there are four possible excited p-wave D_s states which might contribute. These are the two narrow states with the light degrees of freedom in a $j^P = 3/2^+$ state and the two broad states with light degrees of freedom in a $j^P = 1/2^+$ state. Since the states with $j^P = 3/2^+$ decay via d-wave to D^*K_s, they are suppressed. This is also why they are narrow and observable. Of the states with light degrees of freedom in $j^P = 1/2^+$ states, only the 1^+ state contributes. The 0^+ state is forbidden to decay to the final state D^*K_s.

To estimate the above contribution and to calculate the non-resonant amplitude, we use heavy hadron chiral perturbation theory (HHCHPT) [7]. The momentum p_k of K_s can have a maximum value of about 1 GeV for $B^0 \to D^{*+}\bar{D}^{*-}K_s$. This is of the same order as Λ_χ which sets the scale below which we expect HHCHPT to be valid. It follows that in the present case it is reasonable to apply HHCHPT to calculate the three body decays.

In the lowest order in the HHCHPT expansion, contributions to the decay amplitude come from the contact interaction terms and the pole diagrams which give rise to the non-resonant and resonant contributions respectively. The pole diagrams get contributions from the various multiplets involving D_s type resonances as mentioned above. In the framework of HHCHPT, the ground state heavy meson has the light degrees of freedom in a spin-parity state $j^P = \frac{1}{2}^-$, corresponding to the usual pseudoscalar-vector meson doublet with $J^P = (0^-, 1^-)$. The first excited state involves a p-wave excitation, in which the light degrees of freedom have $j^P = \frac{1}{2}^+$ or $\frac{3}{2}^+$. In the latter case we have a heavy doublet with $J^P = (1^+, 2^+)$. These states can probably be identified with $D_{s1}(2536)$ and $D_{sJ}(2573)$ [8]. Heavy quark symmetry rules out any pseudoscalar coupling of this doublet to the ground state at lowest order in the chiral expansion [9]; hence the effects of these states will be suppressed and we will ignore them in our analysis. In fact there is an experimental upper limit on inclusive $B \to D_{s1}(2536)X < 0.95\%$ at 90% C.L [10]. Since the total $D^*\bar{D}^*K$ rate is about 8%, this confirms that the narrow p-wave states do not account for a significant fraction of the total $D^*\bar{D}^*K$ rate.

The other excited doublet has $J^P = (0^+, 1^+)$. Neither of these states has yet been observed even in the D system, because they are expected to decay rapidly through s-wave pion emission and have large widths [11]. Only the 1^+ can contribute in this case. For later reference, we denote this state by $D_{s1}^{*'}$. However, quark model estimates suggest [12] that these states should have masses near $m + \delta m$ with $\delta m = 500$ MeV, where m is the mass of the lowest multiplet.

We will assume that the leading order terms in HHCHPT give the dominant contribution to the decay amplitude and so we will neglect all sub-leading effects suppressed by $1/\Lambda_\chi$ and $1/m$, where m is the heavy quark mass. We show that from the time dependent analysis of $B^0(t) \to D^{*+}\bar{D}^{*-}K_s$ one can extract $\sin(2\beta)$ and $\cos(2\beta)$. Measurement of both $\sin(2\beta)$ and $\cos(2\beta)$ can resolve the $\beta \to \pi/2 - \beta$ ambiguity. We also point out that from the differential distribution of the time independent process $B^0 \to D^{*+}\bar{D}^{*-}K_s$ one can discover the 1^+ resonance $D_{s1}^{*'}$.

EXTRACTION OF β

In this section we discuss the extraction of $\sin 2\beta$ and $\cos 2\beta$ from the time dependent rate for $B(t) \to D^{*+}D^{*-}K_s$.

We define the following amplitudes

$$a^{\lambda_1,\lambda_2} \equiv A(B^0(p) \to D_{\lambda_1}^{+*}(p_+)D_{\lambda_2}^{-*}(p_-)K_s(p_k)), \tag{1}$$

$$\bar{a}^{\lambda_1,\lambda_2} \equiv A(\bar{B}^0(p) \to D_{\lambda_1}^{+*}(p_+)D_{\lambda_2}^{-*}(p_-)K_s(p_k)), \tag{2}$$

where B^0 and $\bar{B}^0$ represent unmixed neutral B and λ_1 and λ_2 are the polarization indices of the D^{*+} and D^{*-} respectively.

The time-dependent amplitudes for an oscillating state $B^0(t)$ which has been tagged as a B^0 meson at time $t = 0$ is given by,

$$A^{\lambda_1,\lambda_2}(t) = a^{\lambda_1,\lambda_2} \cos\left(\frac{\Delta m\, t}{2}\right) + ie^{-2i\beta}\bar{a}^{\lambda_1,\lambda_2} \sin\left(\frac{\Delta m\, t}{2}\right), \tag{3}$$

and the time-dependent amplitude squared summed over polarizations and integrated over the phase space angles is:

$$|A(s^+,s^-;t)|^2 = \frac{1}{2}\Big[G_0(s^+,s^-) + G_c(s^+,s^-)\cos\Delta m\, t - G_s(s^+,s^-)\sin\Delta m\, t\Big], \quad (4)$$

with

$$G_0(s^+,s^-) = |a(s^+,s^-)|^2 + |\bar{a}(s^+,s^-)|^2, \quad (5)$$
$$G_c(s^+,s^-) = |a(s^+,s^-)|^2 - |\bar{a}(s^+,s^-)|^2, \quad (6)$$
$$G_s(s^+,s^-) = 2\Im\left(e^{-2i\beta}\bar{a}(s^+,s^-)a^*(s^+,s^-)\right)$$
$$= -2\sin(2\beta)\,\Re(\bar{a}a^*) + 2\cos(2\beta)\,\Im(\bar{a}a^*). \quad (7)$$

The variables s^+ and s^- are the Dalitz plot variable

$$s^+ = (p_+ + p_k)^2, \quad s^- = (p_- + p_k)^2.$$

The transformation defining the CP-conjugate channel $\bar{B}^0(t) \to D^{-*}D^{+*}K_s$ is $s^+ \leftrightarrow s^-$, $a \leftrightarrow \bar{a}$ and $\beta \to -\beta$. Then:

$$|\bar{A}(s^-,s^+;t)|^2 = \frac{1}{2}\Big[G_0(s^-,s^+) - G_c(s^-,s^+)\cos\Delta m\, t + G_s(s^-,s^+)\sin\Delta m\, t\Big]. \quad (8)$$

Note that for simplicity the $e^{-\Gamma t}$ and constant phase space factors have been omitted in the above equations.

It is convenient in our case to replace the variables s^+ and s^- by the variables y and E_k where E_k is the K_s energy in the rest frame of the B and $y = \cos\theta$ with θ being the angle between the momentum of K_s and D^{*+} in a frame where the two D^* are moving back to back along the z- axis. This frame is boosted with respect to the rest frame of the B with $\vec{\beta} = -(\vec{p}_k/m_B)(1/(1 - E_k/m_B))$. Note $s^+ \leftrightarrow s^-$ corresponds to $y \leftrightarrow -y$. where E'_+ and E_+ are the energy of the D^{*+} in the rest frame of the B and in the boosted frame while E_B is the energy of the B in the boosted frame. The magnitudes of the momentum of the B and the D^{*+} in the boosted frame are given by $|\vec{p}_B|$ and $|\vec{p}_+|$ respectively.

If we neglect the penguin contributions to the amplitude then there is no direct CP violation. This leads to the relation

$$a^{\lambda_1,\lambda_2}(\vec{p}_{k1}, E_k) = \bar{a}^{-\lambda_1,-\lambda_2}(-\vec{p}_{k1}, E_k), \quad (9)$$

where $\vec{p}_{k1}$ is the momentum of the of the K_s in the boosted frame. The above relations then leads to

$$G_0(-y, E_k) = G_0(y, E_k), \quad (10)$$
$$G_c(-y, E_k) = -G_c(y, E_k), \quad (11)$$
$$G_{s1}(-y, E_k) = G_{s1}(y, E_k), \quad (12)$$
$$G_{s2}(-y, E_k) = -G_{s2}(y, E_k), \quad (13)$$

where we have defined

$$G_{s1}(y, E_k) = \Re(\bar{a}a^*), \tag{14}$$
$$G_{s2}(-y, E_k) = \Im(\bar{a}a^*). \tag{15}$$

Carrying out the integration over the phase space variables y and E_k one gets the following expressions for the time-dependent total rates for $B^0(t) \to D^{*+}D^{*-}K_s$ and the CP conjugate process

$$\Gamma(t) = \frac{1}{2}[I_1 + 2\sin(2\beta)\sin(\Delta mt)I_3], \tag{16}$$

$$\overline{\Gamma}(t) = \frac{1}{2}[I_1 - 2\sin(2\beta)\sin(\Delta mt)I_3], \tag{17}$$

where I_1 and I_3 are the integrated G_0 and G_{s1} functions. One can then extract $\sin(2\beta)$ from the rate asymmetry

$$\frac{\Gamma(t) - \overline{\Gamma}(t)}{\Gamma(t) + \overline{\Gamma}(t)} = D\sin(2\beta)\sin(\Delta mt), \tag{18}$$

where

$$D = \frac{2I_3}{I_1}, \tag{19}$$

is the dilution factor.

The $\cos(2\beta)$ term can be probed by by integrating over half the range of the variable y which can be taken for instance to be $y \geq 0$. Measurement of the $\cos(2\beta)$ can resolve the $\beta \to \frac{\pi}{2} - \beta$ ambiguity.

RESULTS

The details of the calculations of the amplitudes and rates using an effective Hamiltonian are given in Ref [13]. HHCHPT gives the interactions of the ground state and excited D mesons with pseudoscalars in terms of two coupling constants. The strength of the coupling of the type $D_s^* D^* K$ is given by g while the coupling of the type $D_{s1}^{*\prime} D^* K$ is given by h.

As inputs to the calculation, we use $f_{D^*} \approx f_{D_s^*} \approx f_{D_{s1}^{*\prime}} = 200$ MeV and take the mass of the $D_{s1}^{*\prime}$ state to be 2.6 GeV. For the Isgur-Wise function we use the form

$$\xi(\omega) = \left(\frac{2}{1+\omega}\right)^2.$$

QCD sum rules have been used to compute the strong coupling constants g and h [14]. We will use $g = 0.3$ as obtained in Ref [14] but keep h as a free parameter because this coupling plays a more important role in the decay widths.

In Fig. 1 we show the branching fraction for $\bar{B}^0 \to D^{*+}D^{*-}K_s$ as a function of the coupling h. A QCD sum rule calculation gives $h \sim -0.5$ [14]. We use the same sign of h as obtained in QCD sum rule calculation but vary h from -0.6 to -0.1. For this range

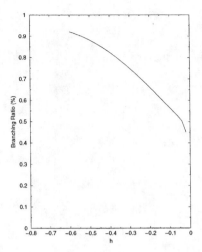

FIGURE 1. The branching fraction for $\bar{B}^0 \to D^{*+}D^{*-}K_s$ as a function of the h.

of h the branching fraction can vary in the range $0.45 - 0.93\%$. For $h = -0.4$ which corresponds to a $D_{s1}^{*\prime}$ state with a width of about 150 MeV the branching fraction is 0.83%. In our calculation this corresponds to a branching ratio $\mathcal{B}(B^0 \to D^{*-}D^{*+}K^0) \approx \mathcal{B}(B^0 \to D^{*-}D^{*0}K^+) \approx \mathcal{B}(B^+ \to \bar{D}^{*0}D^{*0}K^+) \approx \mathcal{B}(B^+ \to \bar{D}^{*0}D^{*+}K^0) \approx 0.9 - 1.86\%$. This is consistent with the CLEO measurements mentioned above.

In Fig. 2 we show a plot of the dilution factor D versus the coupling h. As expected a larger value of $|h|$ gives a larger value of D and hence less dilution in the asymmetry because for a broad $D_{s1}^{*\prime}$ state there is more overlap between the amplitude for $B^0 \to D^{*+}D^{*-}K_s$ and $\bar{B}^0 \to D^{*+}D^{*-}K_s$. For $h = -0.4$ the dilution factor is about 0.75.

In Fig. 3 we show the the decay distribution $d\Gamma/dE_k$ versus the kaon energy E_k. For small values of E_k the decay distribution shows a clear resonant structure which comes from the pole contribution to the amplitude with the excited $J^P = 1^+$ intermediate state. Therefore, examination of the D^*K_s mass spectrum may be the best experimental way to find the broad 1^+ p-wave D_s meson and measure its mass and coupling.

In summary, we have studied the possibility of extracting $\sin(2\beta)$ and $\cos(2\beta)$ from time dependent $B^0 \to D^{(*)}\overline{D}^{(*)}K_s$ decays. These decays are expected to have less penguin contamination and much larger branching fractions than the two body modes $B^0 \to D^{(*)}\overline{D}^{(*)}$. Using HHCHPT we have calculated the branching fractions and the various coefficient functions that appear in the time dependent rate for $B^0 \to D^{(*+)}D^{(*-)}K_s$. We also showed that a examination of the D^*K_s mass spectrum may be the best experimental way to find the broad 1^+ p-wave D_s meson and measure its mass and coupling.

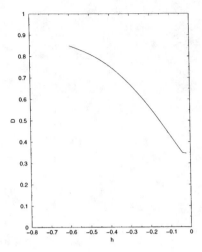

FIGURE 2. The Dilution factor for as a function of the h.

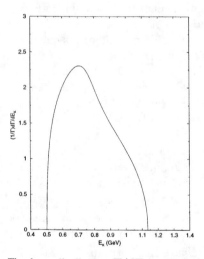

FIGURE 3. The decay distribution $d\Gamma/dE_k$ versus the kaon energy E_k.

ACKNOWLEDGEMENTS

This work was supported in part by the United States Department of Energy (T. Browder and S. Pakvasa), and by the Natural Sciences and Engineering Council of Canada (A. Datta and P. J. O'Donnell).

REFERENCES

1. A. B. Carter and A. I. Sanda, Phys. Rev.**D 23**, 1567 (1981); I. I. Bigi and A. I. Sanda, Nucl. Phys. **B 193**, 85 (1981).
2. I. Dunietz, H. Quinn, A. Snyder, W. Toki and H.J. Lipkin, Phys. Rev. **D 43**, 2193 (1993).
3. G. Michelon, SLAC-BABAR-NOTE-342, Dec 1996.
4. J. Rosner, Phys. Rev. **D 42**, 3732 (1990).
5. CLEO Collaboration, CLEO CONF 97-26.
6. M. Artuso et al. (CLEO Collaboration), Phys. Rev. Lett. **82**, 3020 (1999).
7. M.B. Wise, Phys. Rev. D45 2188 (1992) ; G. Burdman and J. Donoghue, Phys. Lett. **B 280** 287 (1992); T. M. Yan *et al.*, Phys. Rev. **D 46** 1148 (1992); R. Casalbuoni, A. Deandrea, N. Di Bartolomeo, F. Ferruccio, R. Gatto and G. Nardulli, Phys. Rep. **281**, 145 (1997).
8. G. Caso et al., European Journal of Physics, 3, 521 (1998).
9. A. F. Falk and M. Luke, Phys. Lett. **B 292**, 119 (1992).
10. M. Bishai et al (CLEO Collaboration), Phys. Rev. **D 57**, 3847 (1998).
11. N. Isgur and M. B. Wise, Phys. Rev. Lett. **66**, 1130 (1991).
12. S. Godfrey and N. Isgur, Phys. Rev. **D 32**, (1985) 189; S. Godfrey and R. Kokoski, Phys. Rev. **D 43**, 1679 (1991).
13. T. E. Browder, A. Datta, P. J. O'Donnell and S. Pakvasa, hep-ph/9905425.
14. P. Colangelo, A. Deandrea, N. Di Bartolomeo, F. Feruglio, R. Gatto and G. Nardulli, Phys. Lett. **B 339**, 151 (1994); P. Colangelo, F. De Fazio, N. Di Bartolomeo, R. Gatto, G. Nardulli, Phys. Rev. **D 52**, 6422 (1995); P. Colangelo and F. De Fazio, Eur. Phys. J. **C 4**, 503 (1998); V.M. Belyaev, V.M. Braun, A. Khodjamirian and R. Rückl, Phys. Rev. **D 51**, 6177 (1995); S. Narison and H.G. Dosch, Phys. Lett. **B 368**, 163 (1996).

Infrared effects in the decay $B \to X_s \ell^+ \ell^-$

Christian W. Bauer

Department of Physics, University of Toronto
60 St. George Street, Toronto, Ontario, Canada M5S 1A7

Abstract. In this talk I present the calculation of nonperturbative corrections to the inclusive decay $B \to X_s \ell^+ \ell^-$ by performing an operator product expansion up to $\mathcal{O}(1/m_b^3)$. I focus on the interpretation of infrared divergences that arise and estimate the size of the contributions from the nonperturbative matrix elements.

Inclusive heavy quark decays are one of the few hadronic processes that can be calculated model independently beyond the parton level. This is mainly due to the observation that physical observables can be related to the imaginary part of the operator product expansion (OPE) of a forward scattering matrix element [1,2]. This can be written schematically as

$$\Gamma \sim \text{Im} \int e^{-iq\cdot x} \langle B | J^\dagger(x) J(0) | B \rangle$$
$$= \langle B | \mathcal{O}_0 + \frac{1}{m_b} \mathcal{O}_1 + \frac{1}{m_b^2} \mathcal{O}_2 + \ldots | B \rangle \quad (1)$$

where $\mathcal{O}_n$ represents a set of local operators of dimension $(3+n)$.

At leading order this expansion reproduces the parton model result and matrix elements of higher dimensional operators, suppressed by the appropriate power of m_b, parameterize the nonperturbative effects due to the fact that the b quark is bound in the B meson.

A variety of inclusive b decay modes have been analyzed including corrections up to $\mathcal{O}(1/m_b^2)$ [3–5]. Recently $\mathcal{O}(1/m_b^3)$ have been included for some of these decays [6–8]. In this talk I want to present these results for the decay $B \to X_s \ell^+ \ell^-$ [8], focusing on the interpretation of some of the logarithmic divergences that arise.

The decay $B \to X_s \ell^+ \ell^-$, though it has not yet been observed [9], has garnered recent interest because of its sensitivity to new physics that is not probed by other standard model processes. The effective Hamiltonian mediating the $b(p_b) \to s(p_s) + \ell^+(p_+) + \ell^-(p_-)$ transition is obtained from integrating out the top quark and the weak bosons. It is given by

$$\mathcal{H}_{eff}(b \to s\ell^+\ell^-) = -4\frac{G_F}{\sqrt{2}}|V_{ts}^*V_{tb}|\sum_{i=1}^{10} C_i(\mu)O_i(\mu) \quad (2)$$

and the operator basis $\{O_i\}$ can be found in the literature [10].

The Wilson coefficients $\{C_i\}$ at the scale $\mu \sim m_b$ are known in the next to leading log approximation [11]. For consistency with the literature I define two effective Wilson coefficients: $C_7^{eff} \equiv C_7 - C_5/3 - C_6$ and C_9^{eff}. The latter contains the operator mixing of O_{1-6} into O_9 as well as the one loop matrix elements of $O_{1-6,9}$ [11,12]. The full analytic expression for C_9^{eff} is quite lengthy and may be found in [12].

For the branching ratio at the parton level one finds, in agreement with previous calculations [5,10],

$$\frac{\mathcal{B}_{\text{parton}}}{\mathcal{B}_0} = -\frac{32}{9}\left(4 + 3\log\left(\frac{4m_\ell^2}{M_B^2}\right)\right) C_7^{eff\,2} + \frac{2}{3}C_{10}^2 + 128\, C_7^{eff}\int_0^{\frac{1}{2}} dx_0 x_0^2 C_9^{eff}(x_0)$$
$$+ \frac{32}{3}\int_0^{\frac{1}{2}} dx_0 \left(3x_0^2 - 4x_0^3\right) |C_9^{eff}(x_0)|^2 \quad (3)$$

where $x_0 \equiv E_0/m_b$ is the rescaled final state parton energy, and $\mathcal{B}_0$ is the normalization factor

$$\mathcal{B}_0 = \mathcal{B}_{sl}\frac{3\alpha^2}{16\pi^2}\frac{|V_{ts}^*V_{tb}|}{|V_{cb}|^2}\frac{1}{f(\hat{m}_c)\kappa(\hat{m}_c)}. \quad (4)$$

Here $\mathcal{B}_{sl}$ is the measured semileptonic branching ratio, $f(\hat{m}_c)$ is the phase space factor for $\Gamma(B \to X_c\ell\bar{\nu})$

$$f(\hat{m}_c) = 1 - 8\hat{m}_c^2 + 8\hat{m}_c^6 - \hat{m}_c^8 - 24\hat{m}_c^4\log(\hat{m}_c), \quad (5)$$

and $\kappa(\hat{m}_c)$ accounts for the $\mathcal{O}(\alpha_s)$ QCD correction and the leading power corrections. The complete expression for $\kappa(\hat{m}_c)$ may be found in [4]. As alluded to above, the analytic form of C_9^{eff} is sufficiently complicated that we must resort to numerical integrations to calculate the total branching ratio (3).

To include nonperturbative corrections to this result an OPE as in (1) is performed and the matrix elements for operators up to dimension six are parameterized as follows. Matrix elements of dimension four operators vanish at leading order in $1/m_b$ [1] and matrix elements of dimension five operators may be parameterized by λ_1 and λ_2 [13]

$$\langle B(v)|\bar{h}_v \Gamma i D_\mu i D_\nu h_v|B(v)\rangle = M_B \text{Tr}\left\{\Gamma P_+\left(\frac{1}{3}\lambda_1(g_{\mu\nu} - v_\mu v_\nu) + \frac{1}{2}\lambda_2 i\sigma_{\mu\nu}\right)P_+\right\}, \quad (6)$$

where $P_+ = \frac{1}{2}(1+\slashed{v})$ projects onto the effective spinor h_v, and Γ is an arbitrary Dirac structure.

Finally, the dimension six operators may be parameterized by the matrix elements of two local operators [14,15]

$$\frac{1}{2M_B}\langle B(v)|\bar{h}_v iD_\alpha iD_\mu iD_\beta h_v|B(v)\rangle = \frac{1}{3}\rho_1 \left(g_{\alpha\beta} - v_\alpha v_\beta\right) v_\mu,$$

$$\frac{1}{2M_B}\langle B(v)|\bar{h}_v iD_\alpha iD_\mu iD_\beta \gamma_\delta \gamma_5 h_v|B(v)\rangle = \frac{1}{2}\rho_2 \, i\epsilon_{\nu\alpha\beta\delta} v^\nu v_\mu \quad (7)$$

and by matrix elements of two time–ordered products

$$\frac{1}{2M_B}\langle B(v)|\bar{h}_v (iD)^2 h_v i \int d^3x \int_{-\infty}^0 dt\, \mathcal{L}_I(x)|B(v)\rangle + h.c. = \frac{\mathcal{T}_1 + 3\mathcal{T}_2}{m_b},$$

$$\frac{1}{2M_B}\langle B(v)|\bar{h}_v \frac{1}{2}(-i\sigma_{\mu\nu})G^{\mu\nu} h_v i \int d^3x \int_{-\infty}^0 dt\, \mathcal{L}_I(x)|B(v)\rangle + h.c. = \frac{\mathcal{T}_3 + 3\mathcal{T}_4}{m_b}. \quad (8)$$

The contributions from $\mathcal{T}_{1-4}$ can most easily be incorporated by making the replacements [14]

$$\lambda_1 \to \lambda_1 + \frac{\mathcal{T}_1 + 3\mathcal{T}_2}{m_b}$$

$$\lambda_2 \to \lambda_2 + \frac{\mathcal{T}_3 + 3\mathcal{T}_4}{3m_b} \quad (9)$$

in the final analytic results. In addition, as will be explained later, there is a contribution to the total rate from the four fermion operator

$$O^{bs}_{(V-A)} = 2\pi^2 \left(\bar{b}\gamma^\mu L s \bar{s} \gamma^\nu L b\right) \left(g_{\mu\nu} - v_\mu v_\nu\right), \quad (10)$$

$L = (1-\gamma^5)$, with the matrix element

$$f_1 = \frac{1}{2M}\langle B|O^{bs}_{(V-A)}|B\rangle \quad (11)$$

Performing the OPE up to third order in the $1/m_b$ expansion and doing the necessary phase space integrals one obtains the differential decay rate. The analytic expression of this differential decay rate is rather lengthy and is will be presented elsewhere [16]. For the purpose of this talk it suffices to present a plot of the individual contributions of the nonperturbative matrix elements to this differential spectrum which is shown in Fig. 1. I have used the values $\lambda_2 = 0.12\,\text{GeV}^2$ as obtained from the mass splitting of the $B - B^*$ mass splitting and $\bar{\Lambda} = 0.39$ GeV, $\lambda_1 = -0.19\,\text{GeV}^2$ [17]. For the matrix elements of the dimension six operators I use the generic size $(0.5\text{GeV})^3$ as suggested by dimensional analysis [1]. One immediately

---

[1] A more detailed estimate of the matrix element of the four fermion operator f_1 can be found in [18].

notices divergences at both endpoints of this spectrum. The divergence at the $\hat{q}^2 \to 0$ endpoint is due to the intermediate photon going on shell and is a well known feature of the decay $B \to X_s \ell^+ \ell^-$. The analytic form of the divergent term is proportional to

$$\frac{1}{\mathcal{B}_0} \frac{d\mathcal{B}}{d\hat{q}^2}\bigg|_{\hat{q}^2 \to 0} \sim \frac{C_7^2}{\hat{q}^2} \left(1 + \frac{\lambda_1 - 9\lambda_2}{2m_b^2} - \frac{11\rho_1 - 27\rho_2}{6m_b^3} + \frac{\mathcal{T}_1 + 3\mathcal{T}_2 - 3(\mathcal{T}_3 + 3\mathcal{T}_4)}{2m_b^3} \right) \quad (12)$$

The term multiplying the $1/\hat{q}^2$ divergence is (up to a constant prefactor) identical to the total decay rate for the rare inclusive decay $B \to X_s \gamma$ [7]. This explicitly verifies the well known fact that this infrared divergent part of the total decay rate can be written as a convolution of the $B \to X_s \gamma$ rate and the distribution function $f_{e \leftarrow \gamma}(z)$, giving the probability for the photon to fragment into a lepton pair,

$$\Gamma^{IR}_{\mathcal{O}_7}(B \to X_s \ell^+ \ell^-; m_l) = \Gamma^{IR}_{\mathcal{O}_7}(B \to X_s \ell^+ \ell^-; \mu) + \int_0^1 dz f_{e \leftarrow \gamma}(z; \mu, m_l) \Gamma(B \to X_s \gamma). \quad (13)$$

where $\mathcal{O}_7$ mediates the $b \to s\gamma$ transition, z is the fraction of the photon momentum carried by one of the leptons and μ is the factorization scale that separates the long and short distance physics. Experimental cuts, however, require us to stay away

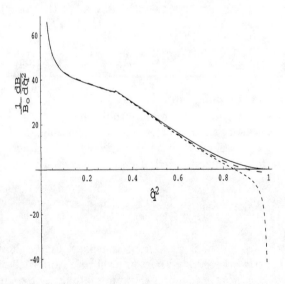

FIGURE 1. The differential decay spectrum. The solid line shows the parton model prediction, the dashed line includes the $1/m_b^2$ corrections and the dotted line contains all corrections including $\mathcal{O}(1/m_b^3)$.

from this endpoint, so by including these cuts in the calculation the divergence will be regulated.

The divergence at the $\hat{q}^2 \to 1$ endpoint is entirely due to the $1/m_b^3$ operators as can be seen from Fig. 1

$$\frac{1}{\mathcal{B}_0} \frac{d\mathcal{B}}{d\hat{q}^2}\bigg|_{\hat{q}^2 \to 1} \sim \frac{\rho_1}{\hat{q}^2 - 1}. \tag{14}$$

This leads to an unphysical logarithmic divergence in the expression for the total rate that is regulated by the mass of the s quark. The existence of such a term has been first noted in [18], where the total decay rate $B \to X_c \ell \bar{\nu}$ was calculated up to $\mathcal{O}(1/m_b^3)$ and it was shown that matrix elements of four fermion operators arise at this order and give contributions proportional to $\rho_1 \log(\hat{m}_c)$. In [19] this was taken a step further and the renormalization group evolution of four fermion operators of dimension six and seven were used to sum phase space logarithms of the form $\hat{m}_c^2 \alpha_s^n \log^n(\hat{m}_c)$, $\hat{m}_c^3 \alpha_s^{n+1} \log^n(\hat{m}_c)$ and $\hat{m}_c^4 \alpha_s^n \log^{n+1}(\hat{m}_c)$. These calculations relied on the fact that the mass of the c quark is large compared to the QCD scale Λ_{QCD} which allowed the c quark to be integrated out of the theory. For the decay $B \to X_s \ell^+ \ell^-$ these methods are not applicable because the s quark is too light. The leading logarithms $\alpha_s^n \log^n(\hat{m}_s)$ are of order unity, making a perturbative calculation of the matrix element of this four fermion operator impossible. This explains why a seventh non-perturbative matrix (11) element is required for this process. It contributes only at the $\hat{q}^2 \to 1$ endpoint of the spectrum and cancels the logarithmic divergence proportional to $\rho_1 \log(\hat{m}_s)$ in the total rate

$$\frac{d\Gamma}{d\hat{q}^2} = \frac{d\Gamma}{d\hat{q}^2}\bigg|_{reg} - 16(C_{10}^2 + (2C_7^{eff} + C_9^{eff})^2) \, \delta\left(q^2 - 1\right) (\rho_1 \log(m_s) - f_1), \tag{15}$$

where $\frac{d\Gamma}{d\hat{q}^2}|_{reg}$ is the function plotted in Fig. 1.

There are additional long distance contributions due to $c\bar{c}$ resonances, which must be cut out before the theory can be compared to measurements. The importance of the nonperturbative corrections when integrated only over a fraction of phase space $\hat{q}^2 > \chi$ should therefore be investigated. Defining a partially integrated branching ratio

$$\mathcal{B}_\chi = \frac{1}{\mathcal{B}_0} \int_\chi^1 d\hat{q}^2 \frac{d\mathcal{B}}{d\hat{q}^2} \tag{16}$$

that depends on the size of the accessible phase space, Fig. 2 shows the fractional correction to this partially integrated parton level rate from each of the nonperturbative parameters λ_i, ρ_i as a function of the minimum accessible dilepton invariant mass χ. In this plot the nonperturbative parameters take the same values as in Fig. 1. Note that the sizes of the ρ_i contributions shown here should not be taken as accurate indications of the actual size of the corrections, but rather as estimates of the uncertainty in the prediction. We see that for $\chi \sim .75$ the contribution from the ρ_1 matrix element is of the same size as the parton model prediction. This is a clear signal that the OPE is no longer valid if the phase space is restricted to be too close to the endpoint $\hat{q}^2 = 1$. This breakdown of the OPE close to the endpoint is a well known fact. Since the CLEO search strategy for this decay imposes the cut $\hat{q}^2 \geq \chi = (m_{\psi'} + 0.1\,\mathrm{GeV})^2/mb^2 = .59$ [9] it is more interesting that even at $\chi \sim .5$ the contribution is about 10%. Investigating this particular value of the cut in more detail we find the individual contributions to the integrated spectrum to be

$$\mathcal{B}_{0.59} = 3.8 + 1.9 \left(\frac{\lambda_1}{m_b^2} + \frac{\mathcal{T}_1 + 3\mathcal{T}_2}{m_b^3} \right) - 134.7 \left(\frac{\lambda_2}{m_b^2} + \frac{\mathcal{T}_1 + 3\mathcal{T}_2}{3m_b^3} \right)$$
$$+ 614.9 \frac{\rho_1}{m_b^3} + 134.7 \frac{\rho_2}{m_b^3} + 560.2 \frac{f_1}{m_b^3}. \quad (17)$$

One can estimate the uncertainty induced by the $\mathcal{O}(1/m_b^3)$ parameters by fixing λ_i to the values given above, then randomly varying the magnitudes of the parameters $\rho_i, \mathcal{T}_i$ and f_1 between $-(0.5\,\mathrm{GeV})^3$ and $(0.5\,\mathrm{GeV})^3$ as suggested by dimensional analysis. We also impose positivity of ρ_1 as indicated by the vacuum saturation approximation [20], and the constraint

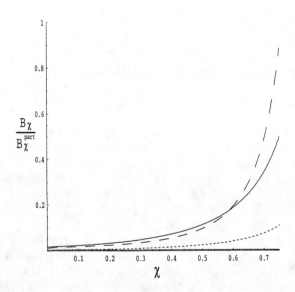

FIGURE 2. The fractional contributions with respect to the parton model result from the higher dimensional operators. The solid, dashed and dotted lines correspond to the contributions from λ_2, ρ_1 and ρ_2, respectively. The contribution from λ_1 is too small to be seen

$$\rho_2 - \mathcal{T}_2 - \mathcal{T}_4 = \left(\frac{\alpha_s(m_c)}{\alpha_s(m_b)}\right)^{3/\beta_0} \frac{M_B^2 \Delta M_B(M_D + \bar{\Lambda}) - M_D^2 \Delta M_D(M_B + \bar{\Lambda})}{M_B + \bar{\Lambda} - \left(\frac{\alpha_s(m_c)}{\alpha_s(m_b)}\right)^{3/\beta_0}(M_D + \bar{\Lambda})} \qquad (18)$$

derived from the ground state meson mass splittings $\Delta M_H = M_{H^*} - M_H$ ($H = B, D$) [14]. Here β_0 is the well known coefficient of the beta function $\beta_0 = 11 - 2/3 n_f$. Taking the 1 σ deviation as a reasonable estimate of the uncertainties from $\mathcal{O}(1/m_b^3)$ contributions to the total rate at this cut, we again find the uncertainty to be at the 10% level. Relaxing the positivity constraint on ρ_1 enlarges the uncertainty to about 20%. Since the cut on q^2 can not be lowered because of the ψ' resonance, these uncertainties are intrinsic to the approach used here.

I would like to thank Craig Burrell for collaboration on this project. This research was funded in part by the Natural Sciences and Engineering Research Council of Canada.

REFERENCES

1. J. Chay, H. Georgi and B. Grinstein, Phys. Lett. **B247** (1990) 399.
2. I.I. Bigi, M. Shifman, N.G. Uraltsev, A. Vainshtein, Phys. Rev. Lett. **71** (1993) 496, B. Blok, L. Koyrakh, M. Shifman, A. Vainshtein, Phys. Rev. **D49** (1994) 3356, Erratum-ibid. **D50** (1994) 3572, A. V. Manohar, M. B. Wise, Phys. Rev. **D49** (1994) 1310.
3. See for expample A.F. Falk, M. Luke and M.J. Savage, Phys. Rev. **D53**, 2491 (1996); Phys. Rev. **D49**, 3367 (1994)
4. A. Ali, L.T.Handoko, G.Hiller, T.Morozumi, Phys. Rev. D **55** (1997) 4105.
5. A. Ali and G. Hiller, Phys. Rev. D **58** (1998) 074001.
6. M. Gremm and A. Kapustin, Phys. Rev. **D55**, 6924 (1997)
7. C. Bauer, Phys. Rev. **D57** (1998) 5611; Erratum, to be published in Phys. Rev **D**.
8. C. W. Bauer and C. N. Burrell, hep-ph yymmnnn
9. S. Glenn et al., Phys. Rev. Lett. **80** (1998) 2289.
10. B.Grinstein, M.Savage, and M.B.Wise, Nucl. Phys. **B319** (1989) 271.
11. M. Misiak, Nucl. Phys. **B393** (1993) 23; **Erratum**, Nucl. Phys. **B439** (1995) 461.
12. A.J. Buras and M.Münz, Phys. Rev. D **52** (1995) 186.
13. A.F. Falk, M.Neubert, Phys. Rev. **D47** (1993) 2965; Phys. Rev. D **47** (1993) 2982.
14. M. Gremm and A. Kapustin, Phys. Rev. **D55** (1997) 6924.
15. T. Mannel, Phys. Rev. **D50** (1994) 428.
16. C.W.Bauer and C.N.Burrell, in preparation.
17. Z.Ligeti, Y.Nir, Phys. Rev. D **49** (1994) 4331; M.Gremm, A.Kapustin, Z.Ligeti, M.Wise, Phys. Rev. Lett. **77** (1996) 20; M.Neubert, Phys. Lett. B **389** (1996) 727.
18. B. Blok, R.D. Dikeman and M. Shifman, Phys. Rev. **D51**, 6167 (1995).
19. C. Bauer, A. Falk and M. Luke, Phys. Rev. D **54** (1996) 2097.
20. M.A. Shifman and M.B. Voloshin, Sov. J. Nucl. Phys. 45 (1987) 292; M.B. Voloshin, N.G. Uraltsev, V.A. Khoze and M.A. Shifman, Sov. J. Nucl. Phys. 46 (1987) 112. Phys. Rev. D **52** (1995) 196.

Predictions for the Semi-Leptonic and Non-Leptonic Decays of the Λ_b and B Meson

Alakabha Datta[1] [a], Harry J. Lipkin [2] [b] and Patrick. J. O'Donnell [3][a]

[a] *Department of Physics,*
University of Toronto, Toronto, Canada.
[b] *Department of Particle Physics,*
Weizmann Institute,
Rehovot 76100, Israel
and
School of Physics and Astronomy,
Tel-Aviv University,
Tel-Aviv 69978, Israel

Abstract. Isospin predictions for the semi-leptonic and non-leptonic decays of the Λ_b baryon are given where isospin conservation of the strong interactions constrains the possible final states in Λ_b decays. Since the baryon hyperfine splittings depend upon light quark dynamics and do not decrease with increasing heavy quark mass, this leads in general to to phase space enhancements in Λ_b decays relative to B meson decays for the same underlying quark transitions making the Λ_b lifetime smaller than the B lifetime. Phase space enhancements in Λ_b decays relative to B decays can be understood in terms of hyperfine interactions in the heavy quark system. The quark-hadron duality for baryons appears to be broken.

INTRODUCTION

The experimental discrepancy between the lifetimes of the Λ_b baryon and the B meson has cast doubts on the validity of the Heavy Quark Effective Theory (HQET) [1]. This suggests a search for effects which can produce a meson-baryon difference and which have not been properly taken into account in HQET [2]. One particular difference is the presence of an isoscalar light quark pair in all isoscalar baryons like the Λ_b and Λ_c consisting of one heavy quark and two light quarks. The masses of these states are lowered by a hyperfine energy due to the interaction between the two light quarks which does not exist in the analogous mesons. This light quark hyperfine energy in baryons does not go to zero at infinite heavy quark mass, in contrast with the meson case and approaches a constant value of order 200 MeV which is by no means negligible in comparison with

---

[1] email: datta@medb.physics.utoronto.ca
[2] email: hjl@hep.anl.gov
[3] email: pat@medb.physics.utoronto.ca. Talk presented by P.J. O'Donnell.

other energy scales relevant to Λ_b decays; e.g. the available kinetic energy that determines the phase space in exclusive decays. Note, for example, that 20% of B decays proceed via the $b \to c\bar{c}s$ transition. The available energies for these decays are of order 1 GeV, only a factor of five above the baryon hyperfine splitting scale.

In this talk [3] predictions coming from isospin invariance and isospin splittings on heavy baryon decays are discussed. We are unable at this stage to provide precise estimates for effects like those discussed above. These depend upon detailed dynamical, spectroscopic and model- dependent information. However our calculations can verify that such effects are serious, and must be taken into account in any detailed model calculation that aims to explain the deviations from HQET.

Isospin conservation is a good approximate symmetry of the strong interactions and can be applied fruitfully in Λ_b decays. The Λ_b baryon is made of a heavy b quark and a ud light diquark system in a spin and isospin singlet state. Semi-leptonic Λ_b decays involve the weak $b \to c$ transition without the involvement of the light quarks. The final hadronic decay products have to be in an isosinglet state as the weak current is an isoscalar. Strong interactions do not change the isospin state of the light diquark which combines with the c quark to form the hadrons in the final state. Single particle hadronic states would therefore dominantly involve the ground and excited Λ_c baryons. In non-leptonic Λ_b decays the effective current×current Hamiltonian gives rise to the following quark diagrams [4] : the internal and external W-emission diagrams, which result in the factorizable contribution, and the W-exchange diagrams which gives rise to the non-factorizable contribution. The W-annihilation diagram is absent in baryon decay and we neglect the penguin contributions. The contribution from the W-exchange diagram is expected to be small in Λ_b decays. The final states in non-leptonic Λ_b decays result from the isosinglet diquark combining with the final state quarks. For instance in the quark level transition $b \to c\bar{c}s$ the diquark can combine with the c quark in the final state. Hence final states like $\overline{D}_s \Lambda_c$ are allowed but states like $\overline{D}_s \Sigma_c$ are not.

As the quark mass becomes heavier many differences among the properties of spin–1/2 and spin–3/2 baryons and also among pseudo scalar and vector mesons containing a heavy quark are expected to become less pronounced [5]. As the quark mass increases it is expected that the lifetimes of particles containing one heavy quark will become very similar [6]. It is in the corrections to the lowest order in Λ_{QCD}/M where models play a role.

In HQET the lifetimes of the Λ_b and the B^0 meson were expected to be the same in the heavy quark limit and just slightly different when certain quark scattering processes that could occur in the Λ_b but not in the meson were included. These principally included (a) the "weak scattering" process, first invoked for the Λ_c^+ lifetime [7], and here of the form, $bu \to cd$, and (b), the so–called "Pauli interference" process $bd \to c\bar{u}dd$ [8,9]. The results of including these terms is a slight enhancement in the decay rate leading to $\tau(\Lambda_b)/\tau(B^0) \sim 0.9$, whereas the evaluation [10] of $\tau(\Lambda_b)$ is 1.24 ± 0.08 ps and $\tau(B^0) = 1.56 \pm 0.04$ ps gives a very much reduced fraction $\tau(\Lambda_b)/\tau(B^0) = 0.79 \pm 0.07$, or conversely a very much enhanced decay rate. There is a recent CDF result [11] which would move this fraction higher than the world average to a value of 0.87 ± 0.11, which is in agreement with both cases.

A possible explanation of the Λ_b lifetime is an enhancement of the decay width, $\Delta\Gamma(\Lambda_b)$ from the $q-q$ scattering. This involves replacing the usual flux factor by $|\psi(0)|^2$, the wave function at the origin of the pair of quarks bu in the Λ_b, (or the pair bd, for which the wave function is the same by isospin symmetry). This wave function at the origin naturally appears in hyperfine splitting [12]. Rosner [13] tried to account for the enhancement by changing the wave function $|\psi(0)_{bu}|^2$; this would also correlate with the surprisingly large hyperfine splitting suggested by the DELPHI group [14]. He was able to show that, under certain assumptions, there could be at most a $13\pm 7\%$ increase of the amount needed to explain the decay rate of the Λ_b. In a more dramatic attempt to explain the lifetime problem it has been proposed [15] to allow the ratio $r = |\psi_{bq}^{\Lambda_b}(0)|^2/|\psi_{b\bar{q}}^{B_q}(0)|^2$ to vary between $1/4$ and 4. Clearly such a large variation is not consistent with the hyperfine relations.

Here we show that isospin conservation leads naturally to a phase space enhancement in Λ_b decays relative to B decays resulting in a shorter lifetime for Λ_b. As shown in [2] isospin conservation chooses the final state in Λ_b decays with the lowest hyperfine energy. In B decays the spectator quark can combine with the c quark to form vector or pseudo scalar final states. The hyperfine energy in this case is averaged out resulting in a phase space advantage for the baryon transition over the meson transition.

In the following sections we study the isospin predictions in semi-leptonic and non-leptonic Λ_b decays. We then show that phase space enhancements in Λ_b decays relative to B decays lead to shorter lifetime for Λ_b relative to B.

SEMI-LEPTONIC DECAYS

Semi-leptonic Λ_b decay involves the quark level $b \to c$ transition due to an isoscalar current. The amplitude for the process can be written as

$$A = <X|J_\mu|\Lambda_b> L^\mu \qquad (1)$$

where $J_\mu = \bar{c}\gamma_\mu(1-\gamma_5)b$ is the isoscalar weak current and L^μ is the leptonic weak current. The final state X has to be in an isosinglet state. In the heavy quark limit the light degrees of freedom in a hadron, the diquark in this case, have conserved isospin and angular momentum quantum numbers. Due to isospin conservation the light diquark in Λ_b remains in an isosinglet state as it combines with the c quark to generate the spectrum of final states. When the diquark combines with the c quark it will form dominantly a Λ type charmed baryon. The Λ type baryon can be classified according to the quantum numbers carried by the light degrees of freedom. So the lowest state corresponds to the light degree having isospin $I_l = 0$ and spin $s_l = 0$. The diquark can be excited to a higher orbital angular momentum state with $L_l = 1$. This creates baryons with net spin $1/2$ and $3/2$ denoted by Λ_{c1}^* and Λ_{c2}^* or alternately as $\Lambda_c(2593)$ and $\Lambda_c(2625)$. Other final states that can populate the X spectrum to a lesser fraction than single particle Λ_c states are $D^0 p$ $D^0 p \pi^0$ etc. Note isosinglet combinations like $\Sigma_c^{++}\pi^-$, $\Sigma_c^+\pi^0$ can only be the decay product of excited Λ_c type baryons where the pion is emitted from the diquark changing it from an isosinglet to an isovector state.

Hence the decay $\Lambda_b \to Xl\bar{\nu}$ should be dominantly $\Lambda_b \to \Lambda^{(*)}l\bar{\nu}$ where $\Lambda^{(*)}$ denotes the ground state or the excited Λ_c. Because the excited Λ_c decays to the ground state Λ_c we have the prediction

$$\Lambda_b \to Xl\bar{\nu} \approx \Lambda_b \to \Lambda_c Xl\bar{\nu}.$$

(See refs. [3,16] for further details.) Thus if a Σ_c is in the final state it must be associated with a π and further the invariant mass of the $\Sigma_c - \pi$ or the $\Lambda_c \pi \pi$ system must be the mass of the excited Λ_c. Finally our prediction is

$$\Lambda_b \to Xl\bar{\nu} \approx \Lambda_b \to \Lambda_c Xl\bar{\nu} = \Lambda_b \to \Lambda_c l\bar{\nu}, \Sigma_c \pi l\bar{\nu}, \Lambda_c \pi \pi l\bar{\nu}$$

In semi-leptonic B decays the largest fraction of final states will involve the D and the D^* meson. When compared to the dominant decay $\Lambda_b \to \Lambda_c l\bar{\nu}$ there is a phase space advantage in Λ_b decays relative to B decays which results in a shorter lifetime for Λ_b relative to B. We will discuss this more quantitatively in a later section.

NON-LEPTONIC DECAYS

Non-leptonic Λ_b decays proceed through the underlying quark transitions $b \to c\bar{c}s'$ and $b \to c\bar{u}d'$ where $d' = d\cos\theta_c + s\sin\theta_c$ and $s' = -d\sin\theta_c + s\cos\theta_c$. We neglect the $b \to u$ and penguin transitions. Non-leptonic transitions involve the W-emission and the W exchange diagrams. From the study of Λ_b lifetime, the W-exchange contribution relative to the spectator b quark decay rate is of the order $32\pi^2|\psi(0)|^2/m_b^3$. This is of the order unity in the case of charmed baryons [9,17] (which has m_c in place of m_b) and so is much suppressed in the case of Λ_b baryons. Note the wave function at the origin, $\psi(0)$, is approximately same for the charm and bottom system. Hence in non-leptonic Λ_b decays the W-exchange term will be small, unlike in the case of charmed baryons, and factorization is expected to be a good approximation.

We now outline the predictions for Λ_b non-leptonic decays which follow from the conservation of the isospin quantum number of the light diquark in the Λ_b baryon. For the $b \to c\bar{c}s$ transition the effective Hamiltonian is

$$H_W = c_1 \bar{c}b\bar{s}c + c_2 \bar{s}b\bar{c}c \qquad (2)$$

where $c_{1,2}$ are the Wilson's coefficients and we have suppressed the color and Dirac index as well as the $\gamma_\mu(1-\gamma_5)$ factors. Since the W-emission diagram, which is given by the factorization amplitude, is the dominant contribution here we can write the non-leptonic amplitude as

$$A[\Lambda_b \to XX'] = (c_1 + c_2/N_c) < X|\bar{c}b|\Lambda_b >< X'|\bar{s}c|0 > \qquad (3)$$
$$A_s[\Lambda_b \to XX'] = (c_2 + c_1/N_c) < X|\bar{s}b|\Lambda_b >< X'|\bar{c}c|0 > \qquad (4)$$

where A and A_s are the color allowed and color suppressed amplitudes and N_c is the number of colors. Now for the color allowed transition, from our analysis of the semi-leptonic decays, X is mainly $\Lambda_c^{(*)}$. Hence some possible final states are $\Lambda_c \bar{D}_s$, $\Sigma_c \pi \bar{D}_s$, $\Lambda_c \pi \pi \bar{D}_s$. Note no single $\Sigma_c(\Sigma_c^*)$ is possible in the final state unless accompanied by a pion. For the color suppressed transitions we can have final states like $\Lambda^{(*)}J/\psi(D\overline{D})$, $\Sigma(\Sigma^*)\pi J/\psi(D\overline{D})$. Again a single $\Sigma(\Sigma^*)$ final state is not allowed unless accompanied by a pion. Also most of the time the $\Sigma\pi$ system would be the decay product of an excited Λ and therefore would have an invariant mass of an excited Λ.

In B decays the same Hamiltonian would generate $D(D^*)\overline{D}_s$ final states in color allowed transitions and and as in the semi-leptonic case when we sum over all $|X'>$ states there will be an enhancement in the Λ_b width relative to the B width. See ref. [3] for a number of specific decay modes.

Λ_b LIFETIME

Lifetimes of the Λ_b and B are calculated using the operator product expansion (O.P.E) to write the square of the decay amplitude as a series of local operators [15]. The expression for the lifetime can be arranged as an expansion in $1/m_b$. The inclusive rate calculated in this manner is expected to equal the inclusive rate by summing up individual exclusive modes by assumption of duality. The validity of duality has not been proved but it can be shown in a certain kinematic limit, the Shifman-Voloshin limit, defined by $m_b, m_c \gg (m_b - m_c) \gg \Lambda_{QCD}$ that the inclusive rate calculated by the method of OPE gives the same result as summing up the exclusive modes which are saturated by $B \to D + D^*$ in B decays and $\Lambda_b \to \Lambda_c$ in Λ_b decays [20,21].

As mentioned in the introduction, in the leading order, the lifetimes of Λ_b and B are expected to be same if the OPE method is used in calculating the lifetimes. Spectator effects that distinguish between Λ_b and B only arise at order $1/m_b^3$ and are not enough to explain the observed Λ_b, B lifetime ratio. In our analysis of exclusive semi-leptonic decays we found that Λ_b goes dominantly to Λ_c while B goes to D and D^*. The result is a phase space advantage in the baryon transition over the meson transition leading to an enhanced Λ_b lifetime relative to the B lifetime. We can calculate the inclusive rates by summing up the exclusive modes. For a quantitative estimate we use the following toy model for semileptonic decays: We assume that the Λ_b goes only to Λ_c, that the B goes to a statistical mixture (3/4) D* and (1/4) D and that all transitions to higher states are small. In the SV limit, in the leading order, for semi-leptonic transition $H_1 \to H_2 l\overline{\nu}$ the decay rates go as $(H_1 - H_2)^5$. In our toy model we will assume that this behavior of the decay rate persists away from the SV limit also. Therefore the phase space for the Λ_b decay is then given by the mass difference $\Lambda_b - \Lambda_c$ to the fifth power. The phase space for the B decay is then given by the $B - D^*$ mass difference to the fifth power, weighted by a statistical factor of (3/4) plus the $B - D$ mass difference to the fifth power, weighted by a statistical factor of (1/4). It is interesting to note that in this toy model $\Gamma(B \to Dl\overline{\nu})/\Gamma(B \to D^*l\overline{\nu}) = 0.41$ which is very close to the experimental number 0.42 [10].

Including small corrections from neglected transitions we can write

$$\Gamma_{SL}(\Lambda_b) = A(\Lambda_b - \Lambda_c)^5(1 + x_1) \tag{5}$$

$$\Gamma_{SL}(B) = A(\frac{1}{4}(B - D)^5 + \frac{3}{4}(B - D^*)^5)(1 + x_2) \tag{6}$$

where A is a constant involving the Fermi constant G_F and $x_{1,2}$ are small corrections from neglected transitions. This well-defined model for semi-leptonic decays may be right or wrong, but its predictions are easily calculated and the basic assumptions can be easily tested when exclusive branching ratios into baryon final states including spin-excited baryons become available. We immediately obtain the following result for the ratio of semi-leptonic partial widths for $x_1 \approx x_2$:

$$\frac{\Gamma(\Lambda_b)}{\Gamma(B)} = 1.07 \tag{7}$$

In a toy model including only semi-leptonic modes this would give the ratio of the life-times

$$\frac{\tau(\Lambda_b)}{\tau(B)} = 0.934 \tag{8}$$

This shows a clear prediction of a significant enhancement of the Λ_b partial semi-leptonic width in comparison with the B. The Λ_b decay rate is enhanced by about 7%.

We see the distinction between the HQET picture of the hadrons and the isospin predictive model clearly arises from the fact that isospin conservation distinguishes between the composition of the other parts of the meson and baryon that do not contain the heavy quark. In contrast to the meson hyperfine mass splittings, which are always between states of the same isospin and decrease to zero with the heavy quark mass, the $\Lambda - \Sigma$ splittings are between states of different isospins and therefore separated from one another by isospin selection rules. Furthermore, they do not decrease with heavy quark mass, but actually increase, and are expected in simple models to approach a finite asymptotic value of 200 MeV with infinite heavy quark mass [22]. In the standard HQET expansion this spin-isospin splitting is neglected and as we have shown above, the effect of the spin-isospin splitting on phase space can be appreciable.

One can also find evidence of a similar enhancement in non-leptonic decays. Consider for instance the quark transition $b \to c\bar{u}d$. Considering only color allowed transitions we found that for Λ_b decays the final states are of the form $\Lambda_c^{(*)} X$ where $X = \pi \rho a_1 n\pi ....$ In the case of B decays the final states are dominantly $D(D^*)X$. If we now sum over the states X then in the leading order we have the effective transitions

$$\Lambda_b \to \Lambda_c^{(*)} \bar{u}d$$
$$B \to D(D^*)\bar{u}d$$

Here we have used the idea of duality in summing over the X states. This maybe reasonable because there are many hadronic channels and so summing over all the final states will eliminate the bound state effects of the individual final states. We can then apply the toy model for semi-leptonic decays considered above and we see that there is a phase space enhancements for Λ_b decays relative to B decays. A similar treatment can also be applied for other color allowed non-leptonic transitions taking proper care of the phase space factors. For instance in $\Lambda_b \to \Lambda_c \bar{u}d$ the invariant mass M_X varies from

$(\Lambda_b - \Lambda_c)$ to $(m_u + m_d)$ and for $\Lambda_b \to \Lambda_c \bar{c} s$ the invariant mass M_X varies from $(\Lambda_b - \Lambda_c)$ to $(m_c + m_s)$.

Note that in the traditional approach to calculating lifetimes using duality the isospin selection rules are not taken into account. For instance the transitions due to $b \to c\bar{c}s$ are $\Delta I = 0$ and so the final states in Λ_b decays are rigorously required to be in an isoscalar state while in B decays only $I = 1/2$ states are allowed. Using duality, in the leading order, both the Λ_b and B decays would be represented by the parton level process $b \to c\bar{c}s$. The dynamics of the two light quarks in the baryon and consequently the fine details of the hadron spectrum is ignored. While it is conceivable that the arguments supporting quark-hadron duality which neglect the fine details of the hadron spectrum maybe valid for mesons it is likely to break down for baryons where there are two valence light quarks undergoing very complicated non-perturbative QCD interactions. We note that the use of duality even in the case of the B meson has been questioned recently [1].

As a concrete example consider the color suppressed $b \to c\bar{c}s$ transition leading to the processes $B \to J/\psi X$ and $\Lambda_b \to J/\psi X$. Possible final states, as already mentioned before, for the Λ_b decay are $J/\psi \Lambda^{(*)}$. In B decays the corresponding final states are in $I = 1/2$ states and some possible final states are $J/\psi K^{(*)}, K\pi\pi$ [10]. If we used duality to sum over the X states then in the leading order both Λ_b and B decaying to final state $J/\psi X$ could be represented by $b \to s J/\psi$ and so the rates for both processes would be same. On the other hand isospin selects specific X states. From measured rates in Particle Data Group [10] $X = \Lambda$ and $K, K^*, K\pi\pi$ for Λ_b and B decays. If we add the observed rates we find $\Gamma[B \to J/\psi X] \sim 6\Gamma[\Lambda_b \to J/\psi X]$. This appears to indicates a breakdown of duality unless there is also significant transition of the Λ_b to excited Λ which could show up as a $\Sigma\pi$ state.

Phase space enhancements in Λ_b decays over B decays can be understood in terms of hyperfine interactions [2,18,19]. In $B \to D^*$ decays there is a phase space disadvantage over $B \to D$ transition because of the higher D^* mass but there is a spin phase space advantage by a factor of three for the D^* in final state over the D in the final state. So the hyperfine energy is averaged out in the $B \to D + D^*$ transition. In Λ_b decays isospin conservation chooses final states with the lowest hyperfine energy. This added hyperfine energy is available for transition and leads to an phase space enhancement in Λ_b decays over B decays (A different argument [23] using the scaling of lifetimes as the inverse fifth power of hadronic rather than quark masses implicitly gives a larger phase space also). Phase space effects were also discussed in a different approach in Ref [24]. The lesson from our analysis is that the effect of phase space enhancements may be a key factor in understanding the lifetime difference between Λ_b and B hadron.

ACKNOWLEDGEMENTS

We would like to thank M. Neubert for useful comments and discussion. This work was supported in part by grant No. I-0304-120-.07/93 from The German-Israeli Foundation for Scientific Research and Development and by the Natural Sciences and Engineering Council of Canada.

REFERENCES

1. N. Isgur, hep-ph/9809279.
2. Harry J. Lipkin and Patrick J. O'Donnell, Phys.Lett. **B409** 412, (1997).
3. This talk is based on the paper by A. Datta, H. J. Lipkin and P. J. O'Donnell, Phys. Lett. **B450**, 250, (1999).
4. L.L. Chau, Phys. Rep. **95**, 1 (1983); L.L. Chau and H.Y. Cheng, Phys. Rev. Lett **56**, 1655; Phys. Rev. **D 36**, 137 (1987).
5. For a review see M. Neubert, Phys. Rep. **245**, 259 (1994).
6. A recent review is by I.Bigi, M. Shifman and N. Uraltsev, hep-ph/9703290, Ann.Rev.Nucl.Part.Sci.**47**, 591 (1997).
7. V. Barger, J. P. Leveille and P. M. Stevenson, Phys. Rev. Lett. **44**, 226 (1980).
8. M . B. Voloshin and M. A. Shifman, Yad. Fiz. **41** 187 (1985) [Sov. J. Nucl. Phys. **41**, 120 (1985)]; Zh. Eksp. Teor. Fiz **91**, 1180 (1986) [Sov. Phys. - JETP **64**, 698 (1986)]; M . B. Voloshin, N. G. Uraltsev, V. A. Khoze and M. A. Shifman, Yad. Fiz. **46**, 181 (1987) [Sov. J. Nucl. Phys. **46**, 112 (1987)].
9. N. Bilić, B. Guberina and J. Trampetić, Nucl. Phys. **B248**, 261 (1984); B. Guberina, R. Ruckl and J. Trampetić, Zeit. Phys **C33**, 297 (1986).
10. C. Caso et al, The European Physical Journal**C3**, 1 (1998).
11. M. Paulini, hep-ex/9903002, to appear in Int. Jour. Mod. Phys. A.
12. A. De Rujula, H. Georgi and S. L Glashow, Phys. Rev. **D12**, 147 (1975).
13. J. L. Rosner, Phys. Lett **B379**, 267 (1996).
14. DELPHI Collaboration, DELPHI 95-107.
15. M. Neubert and C.T. Sachrajda, Nucl. Phys. **B 483**, 339 (1997); M. Neubert, talk at B Physics Conference, Hawaii, March 1997.
16. For recent calculations see M.A. Ivanov *et.al.* Phys. Rev. **D56**, 348, (1997); Debrupa Chakraverty *et.al.* Mod.Phys.Lett. **A12** 195 (1997) and references therein.
17. Hai-Yang Cheng, Phys. Lett. **B 289**, 455 (1992).
18. M. Frank and P. J. O'Donnell, Phys. Lett. **B159**, 174 (1985); Zeit. Phys. **C34**, 39 (1987).
19. H. J. Lipkin, Phys. Lett. **B171**, 293 (1986); Phys. Lett **B172**, 242 (1986).
20. M.B. Voloshin and M.A. Shifman, Yad. Fiz. **47**,801 (1998); Sov. J. Nucl. Phys. **47**, 511 (1988).
21. C.Glenn Boyd, B. Grinstein and A.V. Manohar, Phys. Rev. **D 54**, 2081 (1996).
22. A. De Rujula, H. Georgi and S. L. Glashow, Phys. Rev. **D 12**, 147 (1975).
23. G. Altarelli *et. al.*, Phys. Lett. **B382**, 409 (1996).
24. Changhoa Jin, Phys. Rev. **D 56**, 7267 (1997); Phys. Rev. **D 56**, 2928 (1997).

Multiplicities of Gluon and Quark Jets in pQCD

I.M. Dremin

P.N. Lebedev Physical Institute, Moscow 117924, Russia

Abstract. The review of the energy evolution of average multiplicities of quark and gluon jets in the perturbative QCD is given. Higher order terms in the perturbative expansion of equations for the generating functions are found. First and second derivatives of average multiplicities are calculated. Gluon jets have larger mean multiplicity, and it evolves faster. It is shown which quantities are most sensitive to higher order perturbative and non-perturbative corrections. The energy regions, where the corrections to different quantities are important, are defined. The latest experimental data is discussed.

The progress in experimental studies of properties of quark and gluon jets is very impressive [2,1]. Their separation becomes possible due to new elaborated methods and high statistics at Z^0 resonance that allows to choose the jets with different subenergies. Therefore the detailed study of the energy evolution of such parameters of multiplicity distributions of jets as their average multiplicities and widths becomes also possible.

The perturbative QCD provides quite definitive predictions which can be confronted to experiment [3-7]. Previously, the results of NLO and NNLO were known. Now the next-order (3NLO) terms have been calculated using QCD equations for the generating functions. They show the tendency to approximate the data with better accuracy. However, the theoretically predicted *slope of the ratio* of multiplicities in gluon and quark jets is noticeably smaller than its experimental value. Recently, the experimental data about *the ratio of slopes* of average multiplicities in gluon and quark jets has been reported [1], and analytical calculations of it have been done [8].

The importance of studying the slopes stems from the fact that some of them are extremely sensitive (while others are not) to higher order perturbative corrections and to non-perturbative terms in the available energy region as we show below. Thus they provide us with a good chance to learn more about the structure of the perturbation series from experiment.

In the perturbative QCD, the general approach to studying the multiplicity distributions is formulated in the framework of equations for generating functions (see

[9,6] for more details). In particular, two equations for average multiplicities of gluon and quark jets are written as

$$\langle n_G(y)\rangle' = \int dx \gamma_0^2 [K_G^G(x)(\langle n_G(y+\ln x)\rangle + \langle n_G(y+\ln(1-x))\rangle - \langle n_G(y)\rangle)$$
$$+ n_f K_G^F(x)(\langle n_F(y+\ln x)\rangle + \langle n_F(y+\ln(1-x))\rangle - \langle n_G(y)\rangle)], \quad (1)$$

$$\langle n_F(y)\rangle' = \int dx \gamma_0^2 K_F^G(x)(\langle n_G(y+\ln x)\rangle + \langle n_F(y+\ln(1-x))\rangle - \langle n_F(y)\rangle). \quad (2)$$

Herefrom one can learn about the energy evolution of the ratio of multiplicities r and of the QCD anomalous dimension γ (the slope of the logarithm of average multiplicity of a gluon jet) defined as

$$r = \frac{\langle n_G\rangle}{\langle n_F\rangle}, \qquad \gamma = \frac{\langle n_G\rangle'}{\langle n_G\rangle} = (\ln\langle n_G\rangle)'. \quad (3)$$

Here, prime denotes the derivative over the evolution parameter $y = \ln(p\Theta/Q_0)$, p, Θ are the momentum and the initial angular spread of the jet, related to the parton virtuality $Q = p\Theta/2$, Q_0=const, K's are the well known splitting functions, $\langle n_G\rangle$ and $\langle n_F\rangle$ are the average multiplicities in gluon and quark jets, $\langle n_G\rangle'$ is the slope of $\langle n_G\rangle$, n_f is the number of active flavours. The argument of γ_0^2 in the integrals is chosen as $y + \ln x(1-x)$ i.e. the coupling strength at any vertex depends on the transverse momentum of produced partons.

The perturbative expansion of γ and r is written as

$$\gamma = \gamma_0(1 - a_1\gamma_0 - a_2\gamma_0^2 - a_3\gamma_0^3) + O(\gamma_0^5), \quad (4)$$

$$r = r_0(1 - r_1\gamma_0 - r_2\gamma_0^2 - r_3\gamma_0^3) + O(\gamma_0^4), \quad (5)$$

where $\gamma_0 = \sqrt{2N_c\alpha_S/\pi}$, α_S is the strong coupling constant.

The limits of integration over x (the share of the primary energy) in equations for multiplicities used to be chosen equal either to 0 and 1 or to e^{-y} and $1-e^{-y}$. This difference, being negligibly small at high energies y, is quite important at low energies. Moreover, it is of physics significance. With limits equal to e^{-y} and $1-e^{-y}$, the partonic cascade terminates at the perturbative level Q_0. With limits equal to 0 and 1, one extends the cascade into the non-perturbative region with low virtualities of "children" partons $Q_1 \approx xp\Theta/2$ and $Q_2 \approx (1-x)p\Theta/2$ less than $Q_0/2$. This region contributes terms of the order of e^{-y}, power-suppressed in energy. It is not clear whether the equations and LPHD hypothesis are valid down to some Q_0 only or the non-perturbative region can be included as well.

Previously the values of a_i, r_i were known for $i < 3$. Now we have done the calculations for a_3, r_3 (see [6]). The numerical values of a_i, r_i for different number of active flavours n_f and limits 0–1 are given in the Table. The corresponding behaviour of the ratio r in various approximations (NLO, NNLO, 3NLO) has been confronted to experimental data in [7] with the conclusion that the theoretical curve approaches experimental values the closer the higher order terms have been accounted. The comparatively large value of r_2 helps it but poses some problems for the perturbative treatment of slopes as we discuss below. The NNLO and 3NLO terms contribution to mean multiplicities is almost constant at present energies and does not spoil good agreement on energy dependence obtained within NLO approximation.

n_f	r_1	r_2	r_3	a_1	a_2	a_3
3	0.185	0.426	0.189	0.280	- 0.379	0.209
4	0.191	0.468	0.080	0.297	- 0.339	0.162
5	0.198	0.510	-0.041	0.314	- 0.301	0.112

The perturbative approach should work better for the ratio of the slopes of the multiplicities but it is much less reliable to use the perturbative estimates even at Z^0-energy for such quantities as the slope of r or the ratio of slopes of logarithms of multiplicities. The slope r' is extremely sensitive to higher order perturbative corrections. The role of higher order corrections is increased here compared with r because each nth order term proportional to γ_0^n gets an additional factor n in front of it when differentiated, the main constant term disappears and the large ratio r_2/r_1 becomes crucial:

$$r' = Br_0 r_1 \gamma_0^3 \left[1 + \frac{2r_2 \gamma_0}{r_1} + \left(\frac{3r_3}{r_1} + B_1\right)\gamma_0^2 + O(\gamma_0^3)\right], \qquad (6)$$

where the relation $\gamma_0' \approx -B\gamma_0^3(1 + B_1\gamma_0^2)$ has been used with $B = \beta_0/8N_c$; $B_1 = \beta_1/4N_c\beta_0$ (β's are usual coefficients of the QCD β-function). Let us note that each differentiation leads to a factor α_S or γ_0^2, i.e., to terms of higher order. For values of r_1, r_2, r_3 tabulated above ($n_f = 4$) one estimates $2r_2/r_1 \approx 4.9$, $(3r_3/r_1) + B_1 \approx 1.5$. The simplest correction proportional to γ_0 is more than twice larger 1 at energies studied and the next one is about 0.4. Therefore the ever higher order terms should be calculated for the perturbative values of r' to be trusted.

Let us consider the ratios of slopes $r^{(1)}$ and curvatures $r^{(2)}$ defined as

$$r^{(1)} = \frac{\langle n_G \rangle'}{\langle n_F \rangle'}; \quad r^{(2)} = \frac{\langle n_G \rangle''}{\langle n_F \rangle''}. \qquad (7)$$

Their ratios to r can be written as

$$\rho_1 = \frac{r}{r^{(1)}} = 1 - \frac{r'}{\gamma r}, \qquad (8)$$

$$\rho_2 = \frac{r}{r^{(2)}} = 1 - \frac{2\gamma r r' + r r'' - 2r'^2}{(\gamma^2 + \gamma')r^2}. \qquad (9)$$

Since $r' \propto \gamma_0^3$, the asymptotical values of r, $r^{(1)}$, $r^{(2)}$ coincide and should be equal to 2.25. First preasymptotical corrections are very small. They are of the order of γ_0^2 with a small factor in front and contribute about 2–4% at present energies:

$$\rho_1 = 1 - Br_1\gamma_0^2 \approx 1 - 0.07\gamma_0^2, \tag{10}$$

$$\rho_2 = 1 - 2Br_1\gamma_0^2 \approx 1 - 0.14\gamma_0^2. \tag{11}$$

However, this ideal perturbative situation ends up when next terms are calculated. For example, for ρ_1 their numerical values are

$$\rho_1 = 1 - 0.07\gamma_0^2(1 + 5.38\gamma_0 + 4.21\gamma_0^2) \ (n_f = 4). \tag{12}$$

All the correction terms are much smaller than 1 and, at first sight, one can use this formula. However, the numerical factors in the brackets are so large that at present values of $\gamma_0 \approx 0.5$ the subsequent terms are larger than the first one and the sum of the series is unknown.

Let us note that the lowest order $O(\gamma_0)$ perturbative (NLO) corrections to the ratio of slopes $r^{(1)}$ are the same as for r but higher order corrections are smaller because they are negative both in r and ρ_1 which define $r^{(1)} = r/\rho_1$. That is why experimental values of $r^{(1)}$ (ranging from 1.9 to 2.1 in [1]) are very close to values of r calculated in the NLO-approximation and are less sensitive to higher order corrections.

The interesting feature of the eq. (12) is that one is able to get in ρ_1 the terms up to $O(\gamma_0^4)$ while the value of r in the numerator of (8) is known only up to $O(\gamma_0^3)$. It is related to the fact that $r'/\gamma \sim O(\gamma_0^2)$. Therefore, the very notion of the ordering of approximations (NLO, 2NLO, 3NLO etc) becomes rather conventional.

The value of ρ_1 determines also the ratio of the slope of the logarithm of average multiplicity in a quark jet (γ_F) to that of a gluon jet (γ) as

$$\gamma_F = \rho_1\gamma. \tag{13}$$

Our failure to estimate precisely enough ρ_1 implies that we can not perturbatively evaluate the logarithmic slope of quark jets γ_F as well. Thus it is the ratio of the logarithmic slopes ρ_1 but not the ratio of slopes $r^{(1)}$ which is most sensitive to the higher order perturbative terms.

Even deeper cancellation of higher order corrections should be in $r^{(2)}$. It will be interesting to check this prediction.

The similar procedure can be used to study the behaviour of higher-rank moments of the multiplicity distributions (see the review paper [9]) with the help of the same equations for the generating functions. The most spectacular prediction of these studies was about the oscillations of cumulant moments as functions of their rank. Since then, this effect was observed in many experiments. The predictions about widths as well as higher moments obtained have been also recently confronted to experiment in [2] with the conclusion that there is good correspondence between theory and experiment. The new results about the widths in 3NLO approximation can be found in recent paper [10]. The rather large corrections found there pose some problems and ask for further studies.

To conclude, it is shown that the purely perturbative QCD estimates of the anomalous dimension γ and of the ratio of average multiplicities of gluon and quark jets r are quite reliable at the accuracy level about 10–15% where the non-perturbative corrections become important. For the slope of r (i.e., r') and for the ratio of the logarithmic slopes of average multiplicities ρ_1 the perturbative results become reliable only at much higher energies than Z^0. However, the qualitative trends of theoretical corrections show the right way to better agreement with experiment. In particular, the partial compensation of the higher order corrections in the ratio of the slopes of average multiplicities $r^{(1)}$ results in its good fit by the lowest NLO-expression. These conclusions are obtained with corrections of the order of γ_0^3 in γ and r as given by the equations for the generating functions. To evaluate the slopes at Z^0 and below with higher accuracy one should solve the equations numerically.

REFERENCES

1. DELPHI Coll., Abreu, P., et al, Z. Phys. C **70**, 179 (1996).
2. OPAL Coll., Ackerstaff K., et al, Eur. Phys. J. C **1**, 479 (1998).
3. Dremin I.M., and Nechitailo V.A., Mod. Phys. Lett. A **9**, 1471 (1994); JETP Lett. **58** 881 (1993).
4. Dremin I.M., and Hwa R.C., Phys. Lett. B **324**, 477 (1994).
5. Lupia S., and Ochs W., Phys. Lett. B **418**, 214 (1998); Nucl. Phys. Proc. Suppl. **64**, 74 (1998).
6. Capella A., Dremin I.M., Nechitailo V.A., and Tran Thanh Van J., Nucl. Phys. B (to be published).
7. Dremin I.M., and Gary J.W., Phys. Lett. B (1999) (to be published).
8. Dremin I.M., JETP Lett. **68**, 559 (1998).
9. Dremin I.M., Physics-Uspekhi **37**, 715 (1994).
10. Dremin I.M., Lam C.S., and Nechitailo V.A., Phys. Rev. D (to be submitted).

Continuum Background Suppression Using Various Selectors

Marko Milek and Popat M. Patel

Department of Physics, McGill University
Montreal, Quebec, H3A 2T8

Abstract. Continuum events represent an eminent source of background in any e^+e^- experiment. As these have a higher branching ratio than $B\bar{B}$ events (at *BaBar* this ratio is estimated to about 3.5) or $\tau^+\tau^-$ events, efficient continuum background suppression is essential in many analyses.

Using Artificial Neural Networks and the Nearest Neighbor Method we developed several selectors which, based only on the global event shape variables, efficiently tag $B\bar{B}$ events and $\tau^+\tau^-$ events against the continuum background. These selectors could then be combined with the channel specific information in various types of analyses. The study was done using a parametric Monte Carlo.

INTRODUCTION

The problem of selection, which is successfully associating a candidate with one of the competing hypotheses, is crucial to any high energy physics analysis. Typical examples are individual particle tagging, jet analysis, separating hadronic from electro-magnetic showers, and distinguishing signal from background events.

Traditionally, this is done by applying a series of 'cuts' to various quantities that are known to be relevant for the problem at hand. Accepting or rejecting a given hypothesis is based on whether or not a candidate falls within an allowed 'window' in the considered quantities. Cuts (the allowed windows) are empirically determined so as to achieve the best result possible. This is generally a painstaking procedure.

More elaborate methods can be developed to improve the selection process. The inputs are the quantities one would apply the cuts to if the traditional method was used, and the output determines the hypothesis associated with the candidate. In this note we consider two classification methods: artificial neural network and the nearest neighbor method.

Types of Events

From the expected branching fractions for $B\bar{B}$, $\tau^+\tau^-$ and continuum production at the $\Upsilon(4s)$ resonance ($\sigma_{b\bar{b}} = 1.05$nb, $\sigma_{\tau^+\tau^-} = 0.94$nb, $\sigma_{q\bar{q}} = 3.39$nb [1]) we estimate the ratio of the number of $B\bar{B}$ events produced at $BaBar$ to the number of continuum events to be ≈ 0.31. Of a total number of hadronic events collected on resonance, less than a quarter are $B\bar{B}$. Note that the light leptonic events, $\Upsilon(4s) \to e^+e^-$ and $\Upsilon(4s) \to \mu^+\mu^-$, can be distinguished trivially due to the specific topology of two back to back tracks.

It should be realized that more efficient event selection is possible by a more elaborate analysis using partial or full reconstruction of the event. The goal of this study is to examine the ability to select events based only on the shape. The selector will be designed to return either a probability associated with each of the hypotheses or a flag classifying the event.

Global Event Shape Variables

Global event shape variables provide a valuable insight into the topology of the event. Because $B\bar{B}$ events tend to be more spherical in shape, as opposed to the continuum events which are jet-like, event shape variables can be used to distinguish the two. For this study we have chosen to describe the event shape in terms of six features: number of particles, sphericity, aplanarity, thrust and normalized second and fourth Fox-Wolfram moments. These quantities were introduced to study jet shapes and their formal definitions can be found in [2]. Several other variables, like the angle between the sphericity axis and the thrust axis, or the polar angles of the axes were also considered but including them in the input parameters didn't improve the classification. An attempt to treat charged particles in the event separate from the neutrals (separate event variables were calculated for charged particles and the neutrals) was also rejected.

MONTE CARLO EVENTS

A number of events was created using the $BaBar$ ASLUND Monte Carlo. Generators 1, 10 (KORAL B) and 4 were used for $B\bar{B}$, $\tau^+\tau^-$ and continuum events. Six input parameters (as mentioned before) will be used by the selector classifying the events. Figure 1 shows how the event shape variables vary for $B\bar{B}$ events (cross-hatched histograms), $\tau^+\tau^-$ events (blank histograms) and continuum events (singly-lined histograms).

ARTIFICIAL NEURAL NETWORK SELECTORS

This analysis is a direct continuation of a previously published feasibility study [3]. A detailed description of the methods is given there.

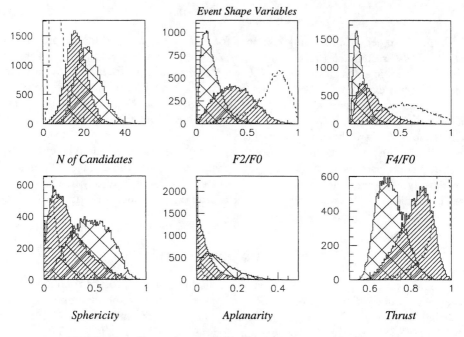

FIGURE 1. Event shape variables for 10,000 events of each type.

Two multi-layered feed-forward artificial neural networks were designed and trained to perform the event selection based on the inputs described. To give equal weight to all six input features they were rescaled to the (-1, 1) range. Several network topologies were studied and 6-5-3-1 (6 neurons in the input layer, two hidden layers with 5 and 3 neurons and a single output neuron) was chosen based on the performance and a small number of network parameters. The number of network parameters should be kept minimal to reduce the computation time during the training phase and to avoid the over-fitting of the training set.

The training events are placed in two disjoint sets. One set is used for the actual training (we use a learning method, called error back-propagation, which iteratively adjusts the network parameters to minimize the deviation between the network output and the desired output on an event by event basis), where the other set is used to judge the convergence of the training procedure. Training is stopped once the error accumulated while running the network on the second set starts to increase. This implies that the training has entered the phase in which particular features of the first set, as opposed to the general properties shared by both sets, are being learned. As we want the network to be able to classify all events which are presented to it, not just the ones belonging to a particular training set, the training must be stopped. One can also stop the training if the desired performance is achieved without the clear convergence (That is, there is no increase in running

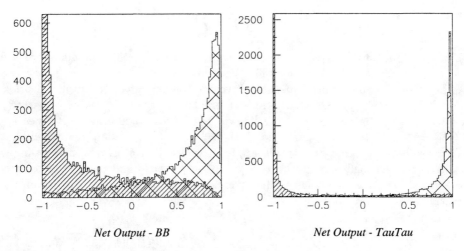

FIGURE 2. Network output for signal (cross-hatched histograms) and background (singly-lined histograms).

error but the error function doesn't decrease further. This happens sometimes if small learning rates are used). In this case there is no danger of over-fitting and no need to train further.

Two separate networks are used to tag $B\bar{B}$ and $\tau^+\tau^-$ events. Each network should output 1 if the events in question is a signal event and -1 for the background events. Note that a network which tags $B\bar{B}$ considers $\tau^+\tau^-$ events as background, and vice versa. Figure 2 is the output of the networks. Signal events (checkered histograms) clearly peak at 1 and background events (crossed histograms) peak at -1. After the training the networks are able to distinguish between signal and background events. Tagging the $\tau^+\tau^-$ events is easier as they differ from the background more significantly than the $B\bar{B}$ events (this can be verified by just looking at the shape variables).

Given two contradicting hypotheses (null hypothesis H_0, and the alternative hypothesis H_1) we would like to obtain a consistency measure which will tell us what is the probability of null hypothesis being true. We will assume that the hypotheses are given in terms of probability density distributions specified by different sets of parameters θ. Network outputs (Figure 2) can easily be converted into probability density functions (p.d.f.). Histograms are smoothed and the parameters θ describing the p.d.f. are just the bin contents. A critical value of the input vector, x_c separates the acceptance region ($x < x_c$) from the rejection region ($x > x_c$).

The probability that the observation x will be rejected when, in fact, it should have been accepted is the *significance*. This is called a *Type I error* and its probability equals α. Another type of error occurs if we accept a null hypothesis when it is false. This, *Type II error* is measured by β and its probability depends on the alternative hypothesis.

$$\alpha = \int_{x_c}^{\infty} f(x|\theta_0)dx \quad \beta = \int_{-\infty}^{x_c} f(x|\theta_1)dx \qquad (1)$$

For a given network output x we need to know what is the probability of the event being signal or background. This depends on the relative values of the two p.d.f.s at the specific x but also on the average frequency (these are just a priori probabilities) of signal versus background. The latter can be estimated, separately for $B\bar{B}$ and $\tau^+\tau^-$ events, from the known $\Upsilon(4s)$ cross sections. The values are $P(\mathrm{H}_0) = 0.1952$ for $B\bar{B}$ signal and $P(\mathrm{H}_0) = 0.1747$ for $\tau^+\tau^-$ signal. Relevant measures of the tagging quality are *efficiency* (fraction of events in the sample which were identified correctly), $error = 1 - efficiency$ and *purity* (fraction of true signal events among the selected events). All of these are dependent on the critical value.

$$\text{efficiency} = P(\mathrm{H}_0)(1-\alpha) + P(\mathrm{H}_1)(1-\beta) \quad \text{purity} = \frac{P(\mathrm{H}_0)(1-\alpha)}{P(\mathrm{H}_0)(1-\alpha) + P(\mathrm{H}_1)\beta} \qquad (2)$$

By varying the value of x_c one can choose either a desired purity, signal efficiency or error in classification (total error is given by $\alpha + \beta$). Dependences of these quantities is shown in Figure 3.

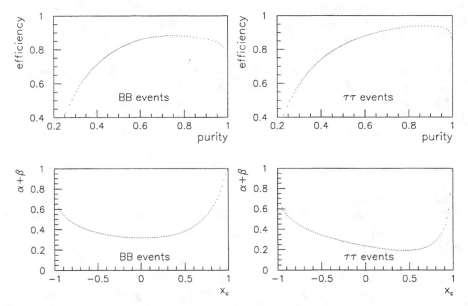

FIGURE 3. Signal efficiency as a function of purity and error as a function of the cut value for $B\bar{B}$ and $\tau^+\tau^-$ events.

To allow simple comparison with other methods which are not easily convertible into a statistic we will define measures of classification quality based on the number of properly and improperly tagged events in the testing set. An event is tagged as signal if the network output is larger then the cut value x_c. As we are interested in detecting signal events, a good tag is a true signal event identified correctly, and a bad tag is a background event identified as signal. We define efficiency ϵ, purity η and dilution $Q = \epsilon \times \eta$ in the following way:

$$\text{efficiency} = \frac{\#\text{good tag}}{\#\text{signal events}} \quad \text{purity} = \frac{\#\text{good tag}}{\#\text{good tag} + \#\text{bad tag}} \quad (3)$$

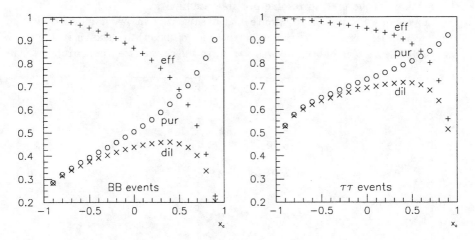

FIGURE 4. Neural network tagging of $B\bar{B}$ and $\tau^+\tau^-$ events as a function of cut value x_c.

NEAREST NEIGHBOR METHOD

A point in an n-dimensional feature space (points represent events and the dimensionality is given by measured parameters, in our case 6 event shape parameters) is classified as belonging to the same class as the point in the training set which is closest to it. Any Minkowski metric can be used to calculate the distance but people tend, for obvious reasons, to be biased toward d_2 which is just the Euclidean metric.

The error of this selection method was proven to be bounded by the optimal Bayes error, $P_e(1 - NN) \leq 2P_e(\text{Bayes})$, yielding a small probability of error. Unfortunately, it is highly impractical to apply it directly as the data structure needed to perform the nearest neighbor search in logarithmic time, $T = \mathcal{O}(\log n)$, is difficult to compute in high dimensions.

Using an algorithm which utilizes a clever data structure based on k-trees, as described in [4], one can find an approximate nearest neighbor in $T = \mathcal{O}(\log n)$ time, which makes the method feasible. The error introduced by the approximation tends to be negligible.

There is no training involved but a large number of parameters is needed to store the information contained in the training set. In principle, all points in the training set should be kept in memory but some data reduction schemes based on proximity graphs tell us how to discard some points (sometimes up to 90% of the set) without a loss in performance. Note that this doesn't reduce the running time considerably.

For better robustness a slightly varied method is used. For an event of an unknown type a sample of k-nearest-neighbors is found, a class appearing the most times in that sample is selected and the event is classified to that class. An average error function can be used to resolve possible ambiguities.

This method was applied to Monte Carlo sets used in previous sections and tagging results (in terms of efficiency, purity and dilution) are shown in Figure 5. A number of nearest neighbors, k, was varied to see its effect on performance. The

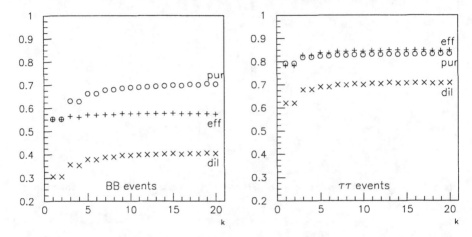

FIGURE 5. Approximate nearest neighbor tagging of $B\bar{B}$ and $\tau^+\tau^-$ events.

results are comparable to those obtained by a neural network but there is no simple way to convert the output, which is just a binary decision stating the belonging to a given class as opposed to the analog output of the neural network, into a statistic and a probability/confidence level for a given class. To accomplish this task one would need to keep track of how many of the k nearest neighbors belonged to either class. Also, in order to introduce a 'cut' parameter for adjusting the desired lever of error versus efficiency one would need to vary the fraction of nearest neighbors needed to make a decision (one could require that more than half of the nearest neighbors belong to a given class if the new event is to be classified). Regardless of the tricks used, running time per event is longer than the running time of the neural network.

To get an estimate of the Bayes probability of error the number of nearest neighbors was increased. The error function of the k-NN method usually has a broad minimum and the $k = \sqrt{n}$, with n being the size of the training set, is generally in the minimum error range. Using $k = 235$ corresponding to the training set of

TABLE 1. Tagging results for all methods used.

	$B\bar{B}$ (ϵ, η, Q)	$\tau^+\tau^-$ (ϵ, η, Q)
F_2/F_0 Cut	0.74, 0.55, 0.41	0.66, 0.81, 0.53
Neural Network	0.74, 0.62, 0.46	0.90, 0.79, 0.72
Nearest Neighbor	0.58, 0.73, 0.42	0.85, 0.83, 0.71

55,000 events we obtained the following values for the efficiency, purity and the dilution factor. For $B\bar{B}$: $\epsilon = 0.58$, $\eta = 0.73$ and $Q = 0.42$, and for $\tau^+\tau^-$: $\epsilon = 0.83$, $\eta = 0.84$ and $Q = 0.70$.

CONCLUSION

We can conclude that all three methods examined yield comparable results for the $B\bar{B}$ and $\tau^+\tau^-$ tagging problem. Moreover, the results are reasonably close to the estimate of the Bayes (optimal) error which was calculated using the k nearest neighbor method.

Traditionally people have been suppressing the continuum background by applying a single cut on the ratio of the second to zeroth Fox-Wolfram moment (F_2/F_0 cut). At CLEO, for instance, a cut value between 0.3 and 0.4 is usually applied to remove a portion of the continuum background from the $B\bar{B}$ sample. Table 1 summarizes the best tagging results (in terms of the highest dilution factor and the corresponding efficiency and purity) for the methods used.

A benefit of the neural network approach is the conversion of the output into a test statistic yielding a consistency with each of the hypotheses. Also, the variation of the critical value x_c allows selection of a sample with given purity, or achieving a pre-determined efficiency. Comparing the neural network selector to standard cuts, an improvement in tagging performance, as measured by the dilution coefficient, of 13% for $B\bar{B}$ events, and of 35% for $\tau\bar{\tau}$ events is observed

Finally, the agreement of both methods with the estimated Bayes error provide a solid consistency check. Further studies using different testing sets could be performed and an estimate of the variance of the selectors could be calculated.

REFERENCES

1. Harrison, H. R., and Quinn, H. R., *The BaBar Physics Book*, SLAC-R-504, 1998, ch. 3, pp. 75.
2. Barlow, R., *Rep. Prog. Phys.* **56**, (1993).
3. Milek, M., and Patel, P. M., *Nucl. Instr. and Methods A* **425/3**, 577 (1999).
4. Mount, D. M., *ANN Programming Manual*, http://www.cs.umd.edu/ mount/ANN/.

New physics in top polarization at the Tevatron

Tibor Torma[1]

Department of Physics, University of Toronto, Toronto, Ontario, Canada M5S 1A7
E_mail: kakukk@physics.utoronto.ca

Abstract. We show that the angular distributions of leptons or jets due to $t\bar{t}$ polarization allow a determination of the top chromomagnetic moment κ with an accuracy of $\mathcal{O}(0.1)$. The method is very stable against background, jet cuts and inaccurate knowledge of each individual event. We propose the inclusion of all-hadronic events for further reduction of the inaccuracies.

I INTRODUCTION

An important goal of high energy experiments is to observe deviations form the Standard Model. Prior to the detection of, or in the absence of, new light resonances such as a Higgs particle, one is left with low energy remnants of loop effects, such as higher dimensional operators. Simple dimensional arguments allow their classification according to their (canonical) dimensions. The lowest dimensional operator is the top chromomagnetic moment, at dimension five,

$$g_s \frac{\kappa}{2m_t} G^a_\mu \bar{t} \frac{\lambda^a}{2} \sigma^{\mu\nu} \partial_\nu t \qquad (1)$$

whose effects in hadron collider experiments have been investigated several times [1,2]. Naively, one would expect that the energy dependence of the new contribution differs significantly from the standard model contribution, as the derivative acting on the top field introduces a momentum factor. Actually the structure of the contributions is schematically as in $1 + \kappa + \kappa^2 \frac{q^2}{m_t^2}$. The would-be enhancement terms $\frac{q}{m_t}$ in the interference term cancel out: the interference between a chirality conserving (the Standard Model) and a chirality changing (κ) amplitude comes by the price of an additional m_t factor. This fact and the unavailability of very large values of q/m_t (most $t\bar{t}$ pairs are produced not far above the threshold) help explain the finding in Ref. [1] that all differential cross

---
[1] Based on work done in collaboration with B. Holdom.

sections in various top momentum variables are insensitive to any $\kappa > 0.1$ other than by an overall change in their normalization. Such an overall normalization factor is hard to measure precisely at the Tevatron because (i) the luminosity is not known with a good accuracy (ii) it is hard to relate an eventual structureless excess of events to this particular operator.

The above discussion leads us to another way to observe the influence of κ. A chirality changing contribution should significantly change the polarization structure of the $t\bar{t}$ state, which results in changes in the angular distributions of the top decay products. Analyses of the $t\bar{t}$ spin structure in the dominant $q\bar{q} \to t\bar{t}$ process are provided in Refs. [3–5]. It has been observed, that in a particularly chosen spin basis dubbed "off-diagonal" [5], each of t and $\bar{t}$ is unpolarized but the polarizations are 100% correlated, up to small corrections from the gluon induced process. Some of the top spin information is carried away by the direction of the charged lepton (or up-type quark), the decay product of the top in the chain $t \to W \to l$. When only this direction is measured in the top rest frame, relative to the top spin direction s, the spin correlation shows up as [5]

$$\frac{1}{\sigma}\frac{d\sigma}{d\cos\vartheta\, d\cos\bar{\vartheta}} = \frac{1}{4}\left[1 + (1 - 2P_\times)\cos\vartheta\cos\bar{\vartheta}\right]. \qquad (2)$$

Here, ϑ and $\bar{\vartheta}$ are angles between s and l on the t and $\bar{t}$ sides respectively, and $P_\times$ is the fractional purity of the unlike-spin component in the $t\bar{t}$ spin density matrix. The latter quantity is minimized by the choice of the off-diagonal spin basis which then provides the maximum observable effect due to top polarization.

Unfortunately, as we will see in the following, this very framework minimizes the effects of the chromomagnetic operator. We have calculated its contribution to $\mathcal{O}(\kappa)$ and found (requiring that both the top quarks and the $W^\pm$ be on shell)

$$\frac{d\sigma(q\bar{q} \to t\bar{t} \to bU\bar{D} + \bar{b}\bar{U}D)}{d\cos\vartheta\, d\varphi\, d\cos\bar{\vartheta}\, d\bar{\varphi}} \propto \mathcal{R}_\kappa \qquad (3)$$

$$\equiv (1 + 2\kappa)(1 + \cos\vartheta\cos\bar{\vartheta})$$
$$-\frac{\beta^2}{2}\sin^2\theta\left[1 + \cos\vartheta\cos\bar{\vartheta} - \sin\vartheta\sin\bar{\vartheta}\cos(\varphi - \bar{\varphi})\right]$$
$$-\gamma\beta^2\kappa\sin\theta\cos\theta\left(\cos\vartheta\sin\bar{\vartheta}\sin\bar{\varphi} + \cos\bar{\vartheta}\sin\vartheta\sin\varphi\right)$$

with $\gamma = (1 - \beta^2)^{-1/2}$ and φ and $\bar{\varphi}$ are the azimuthal angles around the corresponding spin axis s and $\bar{s}$.

We immediately observe that an integration over the azimuthal variables gives

$$\frac{d\sigma}{d\cos\vartheta\, d\cos\bar{\vartheta}} \propto \left(1 + 2\kappa - \frac{\beta^2}{2}\sin^2\theta\right)\left(1 + \cos\vartheta\cos\bar{\vartheta}\right), \qquad (4)$$

so that all information on κ from top polarization is lost. Such is not the situation in any other spin basis where the spin axes $s, \bar{s}$ are turned and the new $\vartheta, \bar{\vartheta}$ angles depend on, in addition to the old $\vartheta, \overline{\vartheta}$, the old azimuthal angles so a factorization like in Eq. (4) does not happen. For this calculational simplicity we continue using the variables defined in the off-diagonal basis.

Eq. (4) actually clarifies the relationship of our approach to both Ref. [1] and [5]. The first factor contains all information used by Rizzo and the 1/2 factor suppresses the dependence on both β and θ, leaving behind only an overall change in the total cross section. The second factor is what has been found by Mahlon and Parke, and that is insensitive to κ.

The complicated structure of the cross section formula presses us to look for a simple variable which still carries most information on κ. However, no simple variable can reproduce the pattern of sign changes in Eq. (3) so is bound to lose much of the information contained in the data.

We are, however, able to answer a simpler question: to find the one best variable that still carries all the information on κ. This variable is provided by the maximum likelihood method which is the standard method of extracting the value of a parameter of a distribution from a sample. We will show that this best variable is (a general statement, valid whenever the rate is linear in κ)

$$F = \frac{\partial_\kappa \mathcal{R}}{\mathcal{R}_{\kappa=0}} \qquad (5)$$

and we find

$$\mathcal{R}_{\kappa=0} = (1 + \cos\vartheta \cos\overline{\vartheta}) - \frac{\beta^2}{2} \sin^2\theta \left[1 + \cos\vartheta \cos\overline{\vartheta} - \sin\vartheta \sin\overline{\vartheta} \cos(\varphi - \overline{\varphi})\right]$$

$$\partial_\kappa \mathcal{R} = 2(1 + \cos\vartheta \cos\overline{\vartheta}) - \gamma\beta^2 \sin\theta \cos\theta \left(\cos\vartheta \sin\overline{\vartheta} \sin\overline{\varphi} + \cos\overline{\vartheta} \sin\vartheta \sin\varphi\right).$$

The maximum likelihood method uses only the values of this variable and integrating out the rest does not affect the determination of κ.

In order to assess the observability of κ in Run 2 of the Tevatron we built a Monte Carlo generator and used a maximum likelihood estimate on the generated data sample to determine the possible statistical accuracy of the estimated value of κ. We found, taking a reasonable 1500 events, a "$1 - \sigma$" accuracy of $\mathcal{O}(\pm 0.1)$. What is quite remarkable is that the accuracy of determination is very stable against factors like (i) the introduction of various cuts, (ii) even crude inaccuracies in the measurement of the parameters of each individual jet, (iii) the inability to tell apart the two hadronic decay products of the W or (iv) the introduction of a background even as large as ten times the $t\bar{t}$ signal. All these insensitivities are due to the complicated pattern of signs in Eq. (5) which is not easily faked or influenced by any of the above factors.

The only factor that has an impact is the number of events. Our numbers show that any sensible determination of κ requires 1000-2000 events and the increase in the event number significantly reduces the inaccuracy. However, for each individual event all that is needed is a very crude estimate of the directions of all six decay products (on the partonic level), which is even possible in the case of all hadronic decay. It seems reasonable to relax the cuts on these events, by the price of larger backgrounds and loss of precision in each event, in order to gain in the event number to provide an estimate of κ which is better than with the determination from the total cross section. In any case, the method advocated here would give an independent cross check which is most welcome when we must tell what the source of a possibly observed cross section anomaly is.

II TOP POLARIZATION AND THE OFF-DIAGONAL BASIS

On the Tevatron the dominating process of $t\bar{t}$ production is $q\bar{q} \to g \to t\bar{t}$. The amplitude of this process at tree level is

$$\mathcal{M}_{t\bar{t}} = \frac{i}{4} \frac{g_s^2}{\hat{s}} \left(\lambda_{q\bar{q}}^a \lambda_{t\bar{t}}^a\right) \left(\bar{v}_{\bar{q}} \gamma^\mu u_q\right) \left[\bar{u}_t \left(1 + \frac{i}{2} \frac{\kappa}{m_t} \sigma^{\mu\nu} k_\nu\right) v_{\bar{t}}\right]. \tag{6}$$

which directly translates into a spin density matrix for the top

$$\sum |\mathcal{M}_{t\bar{t}}|^2 = \frac{16\pi^2 \alpha_s^2}{9} \rho_{s\bar{s},s'\bar{s}'} \tag{7}$$

where color and quark spin summation/averaging is understood. The indeces $s, \bar{s}$ refer to the top (antitop) spin, and $s', \bar{s}'$ are the corresponding spin indeces in the conjugate amplitude. They are all measured with respect to a (for the moment) arbitrary spin direction $s^\mu, \bar{s}^\mu$. For definiteness, we choose the spacial part of the $s^\mu, \bar{s}^\mu$ vectors back-to-back at an angle ψ to the beam direction in the ZMF frame.

The structure of the density matrix is simple in the off-diagonal basis [5]

$$\tan \psi = \frac{\beta^2 s_\theta c_\theta}{1 - \beta^2 s_\theta^2} \quad \text{with} \quad -\frac{\pi}{2} < \psi < \frac{\pi}{2} \tag{8}$$

where standard model contributions from like top and antitop spins vanish:

$$\rho_{off-diag} = \begin{bmatrix} 0 & ig_1 s_\theta c_\theta & -ig_1 s_\theta c_\theta & 0 \\ -ig_1 s_\theta c_\theta & 2(1+2\kappa) - \beta^2 s_\theta^2 & \beta^2 s_\theta^2 & ig_1 s_\theta c_\theta \\ ig_1 s_\theta c_\theta & \beta^2 s_\theta^2 & 2(1+2\kappa) - \beta^2 s_\theta^2 & -ig_1 s_\theta c_\theta \\ 0 & -ig_1 s_\theta c_\theta & ig_1 s_\theta c_\theta & 0 \end{bmatrix} \tag{9}$$

where we used the shorthands $s_\theta = \sin\theta$ and $c_\theta = \cos\theta$. This matrix, in the pure SM limit, reproduces the density matrix found in [5].

The spin correlations encoded in the density matrix will be observed in the angular distributions of the decay products. In the next section we find these distributions for each spin state of the $t\bar{t}$ system and find that all but the diagonal elements in ρ depend on the azimuthal angles φ and/or $\bar{\varphi}$. When we observe only the cross sections integrated in these variables, we only see the diagonal elements of ρ. However, up to normalization, the structure of the diagonal part is identical to that in the standard model case, so we loose all information on κ contained in the spin structure.

III THE SIX PARTICLE PHASE SPACE AND THE DIFFERENTIAL CROSS SECTION

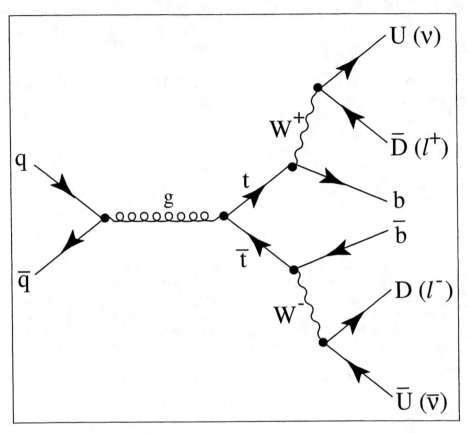

FIGURE 1. The full cascade of $t\bar{t}$ production and decay. The decay products of the W^+ are an up-type $U = u, c$ quark (or a neutrino) and a down-type antiquark $\bar{D} = \bar{d}, \bar{s}$ (or a positively charged lepton, e^+ or μ^+).

The correct choice of the variables in the six particle phase space (see Fig. 1) should keep the expressions for the amplitude simple and in the same time should allow us to write the six-particle phase space measure as simple as possible. We should make sure the regions of integration are also simple. For example, no combination of angles and/or energies in the ZMF frame will satisfy these requirements. For these reasons we have chosen the following set of variables, in addition to the ones that define the ZMF frame:

- The scattering angle θ between q and t in ZMF and the corresponding (irrelevant) azimuthal angle defining the scattering plane.

- The angle ϑ between the spin direction s and the $\overline{D}$ quark (or charged lepton) and the corresponding azimuthal angle φ in the top frame.

- The angle $\overline{\vartheta}$ between the spin direction $\overline{s}$ and the D quark (or charged lepton) and the corresponding azimuthal angle $\overline{\varphi}$ in the antitop frame.

- The angle λ between the direction of W^+ and the $\overline{D}$ quark and the corresponding azimuthal angle ϕ; also the respective angles $\overline{\lambda}, \overline{\phi}$ on the $\overline{t}$ side.

These ten variables completely describe the six-particle phase space (which is 10 dimensional due to four restrictions from momentum conservation and four from the on shell conditions of the $t, \overline{t}, W^\pm$). Their corresponding ranges are completely independent of each other and the phase space measure is expressed in a very simple way. The price to pay is of course that these angles are not all defined in the same frame.

We calculated the rate of the decay of each of the 16 polarization states of the $t\overline{t}$ system, used a Breit-Wigner formula to find the $q\overline{q} \to 6\, quarks$ cross section. The differential cross section is proportional to

$$\mathcal{R}_\kappa = \frac{1}{4} \sum_{s\overline{s} \, s'\overline{s}'} \rho_{s\overline{s},s's'}(\theta) \begin{bmatrix} 1 + \cos\vartheta & \sin\vartheta e^{i\varphi} \\ \sin\vartheta e^{-i\varphi} & 1 - \cos\vartheta \end{bmatrix}_{ss'} \begin{bmatrix} 1 - \cos\overline{\vartheta} & \sin\overline{\vartheta} e^{i\overline{\varphi}} \\ \sin\overline{\vartheta} e^{-i\overline{\varphi}} & 1 + \cos\overline{\vartheta} \end{bmatrix}_{\overline{s}\overline{s}'} \quad (10)$$

Note here that this formula does not make use of the particular choice of the spin basis. In each spin basis the nondiagonal elements of the density matrix are observable only in the azimuthal angles (whose definition of course depends on the chosen basis). The choice of the off-diagonal variables is motivated by the simplicity of the density matrix. Substituting Eq. (9) into Eq. (10) gives $\mathcal{R}_\kappa$ in Eq. (3), so that we managed to express all polarization information in terms of five angles and the top quark velocity.

IV NUMERICAL RESUTS

We estimated κ in samples of various sizes on samples with various true values of κ fed into the MC generator. Following tradition, we will quote the interval where the log-likelihood function drops by $1/2$ as a $1 - \sigma$ interval, although we have to keep in mind that such a statement is mathematically inaccurate. We found that the estimation is consistent with the true value of κ and corresponds to a 85% probability level.

We ran each estimation (and event generator) with three different initial setting of the random generator. We found the following $1 - \sigma$ intervals (we quote the average of the three values):

$$N = 200 \quad \Delta\kappa = \begin{matrix} +0.30 \\ -0.15 \end{matrix}$$
$$N = 900 \quad \Delta\kappa = \begin{matrix} +0.22 \\ -0.12 \end{matrix}$$
$$N = 1500 \quad \Delta\kappa = \begin{matrix} +0.10 \\ -0.07 \end{matrix} \qquad (11)$$
$$N = 2000 \quad \Delta\kappa = \begin{matrix} +0.10 \\ -0.07 \end{matrix}$$
$$N = 500 \quad \Delta\kappa = \begin{matrix} +0.05 \\ -0.03 \end{matrix}$$

We immeditely observe the importance of the event number. The number of observed $t\bar{t}$ events on Run 2 is expected to be $\mathcal{O}(1500)$. With that number the accuracy is in the order of $\Delta\kappa = 0.1$, not very far from the prediction of some models of new physics.

The next question we must address is whether experimental realities significantly deteriorate these results. We included the following issues:

- In the hadronic decay mode of the W it is almost impossible to tell which of the two jets comes from which quark. We set up a similar maximum likelihood estimation of κ where the new probability function now includes a 50% probability of misidentifying the two jets. The change from the values in (11) is completely insignificant. This statement is true for "symmetrization" on only one side (i.e. for the W^+ only, relevant for semileptonic events) and for both sides, relevant for all-hadronic decays.

- We imposed several sets of cuts on the data to see their effect on the extraction. We imposed the same parton level cuts that had been used in the analysis of the semileptonic events in Run 1 [6], (a charged lepton with $E_T > 8\,GeV$, $|\eta| < 1.0$, a neutrino with $E_T > 20\,GeV$, three jets with $E_T > 15\,GeV$ and $|\eta| < 2.0$ and one more jet with $E_T > 8\,GeV$ and $|\eta| < 2.4$. We found no significant change in the accuracy when the number of events passing the cuts were fixed ($\Delta\kappa = \begin{matrix}+0.20\\-0.11\end{matrix}$ for $N = 900$ and $\Delta\kappa = \begin{matrix}+0.13\\-0.08\end{matrix}$ for $N = 1500$). Imposing a jet separation cut of $\Delta\alpha > 20^0$ similarly had little effect.

- The relation of jet observables to the corresponding quark three-momenta introduces errors which are aggravated by the sizable probability of matching some hadrons with the wrong jet. As a first estimate of this effect we set up a random distortion of the energy and direction of each jet by a relative portion of $\delta = 30\%$. Imposing the top and W mass conditions on the data we found again no significant changes ($\Delta\kappa = \begin{matrix}+0.22\\-0.11\end{matrix}$ for $N = 900$ and $\Delta\kappa = \begin{matrix}+0.12\\-0.09\end{matrix}$ for $N = 1500$). Using $\delta = 90\%$ provided a cross-check: we found $\Delta\kappa > 1.0$ so such a heavy distortion of the data is needed to destroy its usefulness.

- Finally, we included a very crude model of the background. Because we expect that the background events, which do not involve a top quark, will not know about the correlation between the t and $\bar{t}$ spins, their distribution is not similar to that of the top events. Consequently, the precise form of the background distribution in unimportant and a uniform distribution can be used in the angular variables. We found, again keeping the number of $t\bar{t}$ events constant $N = 1500$ on top of the background, that the size of the $1 - \sigma$ interval slightly but significantly increases:

$$
\begin{aligned}
N_{total} = 15,000 \quad & N_{t\bar{t}} = 1500 \quad \Delta\kappa = {+0.15 \atop -0.09} \\
N_{total} = 4,500 \quad & N_{t\bar{t}} = 1500 \quad \Delta\kappa = {+0.13 \atop -0.08} \\
N_{total} = 3,000 \quad & N_{t\bar{t}} = 1500 \quad \Delta\kappa = {+0.11 \atop -0.07} \\
N_{total} = 1,875 \quad & N_{t\bar{t}} = 1500 \quad \Delta\kappa = {+0.11 \atop -0.07} \\
N_{total} = 1,500 \quad & N_{t\bar{t}} = 1500 \quad \Delta\kappa = {+0.10 \atop -0.07}
\end{aligned}
\qquad (12)
$$

This slight increase, however, shows only that even a large background can be tolerated in exchange for a larger event number.

In sum, we see that many of the experiment-related issues that usually deteriorate the accuracy of an extraction have little effect in our case, due to the fact they cannot easily mimic the intricacies of the angular distribution in Eq. (5). We expect a similar conclusion to hold for the here uncalculated systematic errors.

ACKNOWLEDGEMENTS

We would like to thank Andrew Robinson for useful discussions. This work was supported by the Natural Sciences and Engineering Research Council of Canada.

REFERENCES

1. D. Atwood, A. Kagan and T.G. Rizzo, Phys. Rev. **D52**, 6264, (1995), hep-ph/9407408
2. K. Cheung, Phys. Rev D55, 4430 (1997), hep-ph 9610368.
3. G. Mahlon and S. Parke, Phys. Rev. **D53**, 4886, (1996), hep-ph/9512264
4. S. Parke and Y. Shadmi, Phys. Lett. **B387**, 199 (1996), hep-ph/9606419
5. G. Mahlon and S. Parke, Phys. Lett. **B411**, 173 (1997), hep-ph/9706304
6. F. Abe et al., Phys. Rev. **D59**, 1 (1999)

Gluons in a Color-Neutral Nucleus

Gregory Mahlon

Department of Physics, McGill University
3600 University St., Montréal, QC, Canada H3A 2T8

Abstract. We improve the McLerran-Venugopalan model [1,2] by introducing a charge-density correlation function which is consistent with the observation that nucleons carry no net color charge. The infrared divergence in the transverse coordinates that was present in the McLerran-Venugoplan model is eliminated by the enforcement of color neutrality.

The problem of extracting first principles predictions from the theory of strong interactions, quantum chromodynamics (QCD), is notoriously difficult, largely because of the nonlinearity of the theory. The quanta of the gauge field, the gluons, themselves carry color charge and so serve as a source of additional gluons. Thus, any regime in which it is possible to actually compute some physical observable within the framework of QCD is of great interest.

Such is the case with the McLerran-Venugopalan (MV) model [1,2]. What McLerran and Venugopalan realized is that for very large nuclei at very small values of the longitudinal momentum fraction, the number of color charges participating in the generation of the QCD vector potential is large. In this situation, the gluon number density may be approximated to lowest order by solving the *classical* Yang-Mills equations in the presence of the (classical) source generated by the valence quarks. To actually extract the gluon number density, we must average over the sources to obtain the two point correlation function for the vector potential. This approximation may be systematically improved by including the quantum corrections. The correlation function derived in Ref. [2] is highly infrared divergent at large transverse distances. In this talk, we will show that this difficulty may be ameliorated by forcing the nucleons to obey a color-neutrality condition [3].

Before proceeding with the main part of this talk, it is necessary to say a few words about our notation and conventions. We elect to work with light-cone coordinates, defined by $x^{\pm} \equiv (x^0 \pm x^3)/\sqrt{2}$. The components of a 4-vector will be written as $x = (x^+, x^-, \vec{x})$, with vectors in the two-dimensional (transverse) subspace written with arrows. We choose a metric with the signature $(-,+,+,+)$. Consequently, we have $x_{\pm} = -x^{\mp}$, and a dot product which reads $x \cdot q \equiv -x^+ q^- - x^- q^+ + \vec{x} \cdot \vec{q}$. We will work in the light-cone gauge, $A^+ \equiv 0$, where the intuitive parton model is

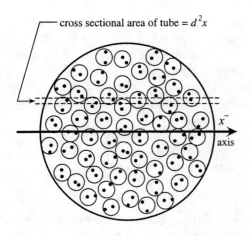

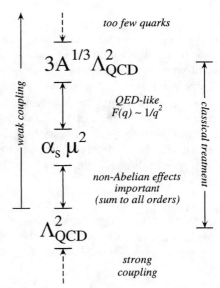

FIGURE 1. A cartoon of a very large nucleus in its rest frame. In the labatory frame, Lorentz contraction causes all of the quarks and nucleons to pile up at essentially the same value of x^-.

FIGURE 2. Map of transverse momentum scales in the McLerran-Venugopalan model.

realized [4].

Within the MV model [1,2], the Lorentz-contracted nucleus is treated as a pancake of color charge coming from the valence quarks. For a large nucleus, this means that a tube of cross-sectional area $d^2\vec{x}$ will intercept a large number of valence quarks (see Fig. 1). If we restrict ourselves to small longitudinal momentum fractions, then all of the quarks in the tube effectively overlap. Because a large number of quarks contribute, the color charge is in a high-dimensional representation of the group, and may be approximated by a classical charge density.

To compute the gluon number density, we begin by solving the classical Yang-Mills equation for the vector potential A^μ as a function of the color charge density ρ. Next, the usual quantum mechanical average is approximated by an average over an ensemble of nuclear charge distributions to compute two-point correlation functions of the vector potential, $\langle A_i^a(x^-, \vec{x}) A_j^b(x'^-, \vec{x}') \rangle$. These correlation functions contain the gluon number operator; hence, they are be connected to the gluon number density. The result of the averaging process depends on the form chosen for the two-point charge density correlation function $\langle \rho^a(x^-, \vec{x}) \rho^b(x'^-, \vec{x}') \rangle$. Finally, we obtain the gluon number density by performing the appropriate Fourier transform of the position space correlator.

Under what conditions is it legitimate to use the classical approximation we have just described? Schematically, the situation is as illustrated in Fig. 2. If the typical transverse momenta being considered is too small (below Λ_{QCD}), then we are in the stong coupling regime, where we have no reason to expect classical methods to be valid. At larger transverse momentum scales, we have weak coupling. However, we cannot go to too large a transverse momentum, or else the requirement that we have a large number of quarks contributing in each patch $d^2\vec{x}$ of nucleus no longer holds: the quantum graininess begins to become important. At the upper end of where the classical approximation is valid, the solution is QED-like, with a distribution function which is proportional to $1/\vec{q}^{\,2}$. Somewhere in between is a cross-over scale where the non-Abelian terms in the equations of motion become important, and we require a solution which is summed to all orders.

We now turn to an examination of the basic form of the classical solution, using the situation in electrodynamics as our guide. Consider a point charge e moving at the speed of light down the z axis. In this situation, the current is

$$J^+ = e\,\delta(x^-)\delta^2(\vec{x}); \quad J^- = 0; \quad \vec{J} = 0. \tag{1}$$

Unlike in the (more familiar) Lorentz gauge where the μth component of the current generates the μth component of the vector potential, in light cone gauge only the transverse components of A^μ are non-zero:

$$A^i(x) = \frac{e}{2\pi}\,\Theta(x^-)\,\frac{x^i}{\vec{x}^{\,2}}. \tag{2}$$

This vector potential corresponds to the non-vanishing field tensor components

$$F^{+i} = -\frac{e}{2\pi}\,\delta(x^-)\,\frac{x^i}{\vec{x}^{\,2}}. \tag{3}$$

These components of the field tensor correspond to transverse electric and magnetic fields of equal strength. There are no longitudinal fields in this limit. An observer sitting at some fixed position $(\vec{b}, x^3)$ would see no fields except at the instant when the charge made its closest approach, when a δ-function pulse would be seen. The magnitude of the pulse would be proportional to $1/b$. The (unobservable) vector potential, which was zero before the passage of the charge, takes on a non-zero value for all times afterward. Since the fields vanish at these times, the late-time vector potential may be thought of as some particular gauge transformation of the vacuum.

The overall features of the above description continue to hold when we switch to QCD, although the details are slightly altered by the presence of the non-Abelian terms in the field equations. The net effect of these terms is to color-rotate the source in a complicated fashion. Nevertheless, the chromoelectric and chromomagnetic fields are non-zero only at the instant of closest approach by the charge, and the vector potential switches from one gauge transform of the vacuum to a different one at this instant.

We now outline the method used to compute the correlation function $\langle A_i^a(x^-,\vec{x})A_j^b(x'^-,\vec{x}')\rangle$. Full details of this part of the calculation are found in Ref. [2]. Essentially, what one does is to expand the vector potential in "powers" of the charge density ρ, and perform the averaging by doing all possible pairwise contractions using

$$\langle \rho^a(x^-,\vec{x})\rho^b(x'^-,\vec{x}')\rangle = \delta^{ab}\,\mu^2(x^-)\,\delta(x^- - x'^-)\,\mathcal{D}(\vec{x}-\vec{x}'). \quad \text{RhoRho} \quad (4)$$

In this expression $\mu^2(x^-)$ is the color charge squared per unit area per unit thickness (*i.e.* per unit x^-). We set up the calculation with a non-zero nuclear thickness to avoid ambiguities in the commutator terms of the Yang-Mills equations which would arise if we let $\rho \sim \delta(x^-)$ exactly. Refering back to Fig. 1, we see that the quarks in our "tube" of color charge typically come from different nucleons. Thus, we expect them to be uncorrelated, hence the dependence $\delta(x^- - x'^-)$ in Eq. (4). The transverse dependence of the charge denisty correlator is given by $\mathcal{D}(\vec{x}-\vec{x}')$. In Ref. [2], it is argued that this should also be a delta-function, since we are restricted to length scales $\lesssim \Lambda_{\text{QCD}}^{-1}$. However, as we shall see, doing so is not consistent with color neutral nucleons, and leads to severe infrared divergences in the correlation funtion.

After doing all of the contractions, we resum the series to obtain the master formula

$$\langle A_i^a(x^-,\vec{x})A_j^b(x'^-,\vec{x}')\rangle = \frac{\delta^{ab}}{N_c}\frac{\partial_i \partial'_j L(\vec{x}-\vec{x}')}{L(\vec{x}-\vec{x}')}\left[e^{N_c \chi(x^-,x'^-)L(\vec{x}-\vec{x}')} - 1\right]. \quad (5)$$

Eq. (5) depends on two new functions. The first, $\chi(x^-,x'^-)$, measures the amount of charge in those layers of the source which have already passed *both* of the points which we are comparing:

$$\chi(x^-,x'^-) \equiv \int_{-\infty}^{\min(x^-,x'^-)} d\xi^-\,\mu^2(\xi^-). \quad \text{ChiDef} \quad (6)$$

The appearance of this function may be understood by recalling that the value of the vector potential depends on whether or not the charge has yet reached its point of closest approach. Although the range of integration in (6) extends to $x^- = -\infty$, in practice this is cut off by the form of $\mu^2(x^-)$, which for a pancake-shaped charge distribution, should be non-zero only in a relatively small range near the value of x^- that corresponds to the position of the nucleus.

The second new function appearing in (5) is given by

$$L(\vec{x}-\vec{x}') \equiv \int d^2\vec{\xi}\int d^2\vec{\xi}'\,\mathcal{D}(\vec{\xi}-\vec{\xi}')\big[G(\vec{x}-\vec{\xi})G(\vec{x}'-\vec{\xi}')$$
$$-\tfrac{1}{2}G(\vec{x}-\vec{\xi})G(\vec{x}-\vec{\xi}') - \tfrac{1}{2}G(\vec{x}'-\vec{\xi})G(\vec{x}'-\vec{\xi}')\big], \quad (7)$$

where

$$G(\vec{x}-\vec{x}') \equiv \frac{1}{4\pi}\ln\left(\frac{|\vec{x}-\vec{x}'|^2}{\lambda^2}\right). \quad (8)$$

This Green's function is the inverse of the operator $\vec{\nabla}^2$ (in two dimensions). The lack of a scale in our theory produces an infrared divergence in this function, which is signalled by the appearance of an arbitrary length scale λ. Clearly, the infrared finiteness or lack thereof of $L(\vec{x} - \vec{x}')$ is intimately related to the form chosen for $\mathcal{D}(\vec{\xi} - \vec{\xi}')$. On the other hand, according to Eq. (7), $L(\vec{x} - \vec{x}')$ vanishes when $\vec{x} = \vec{x}'$, *i.e.* in the ultraviolet. Thus, at very short distances the nonlinear terms in (5) drop out and the behavior of the correlation function is the same as if we had considered a purely Abelian theory instead.

If, as in Ref. [2], we take $\mathcal{D}(\vec{\xi} - \vec{\xi}') = \delta^2(\vec{\xi} - \vec{\xi}')$, we end up with

$$L(\vec{x} - \vec{x}') \sim \frac{1}{\vec{\nabla}^4}\left[\delta^2(\vec{x} - \vec{x}') - \delta^2(0)\right]. \tag{9}$$

Although the subtraction term serves to remove the quadratic infrared singularity, a logarithmic divergence remains. Assuming that the arbitrary scale should be of order Λ_{QCD} on physical grounds, the authors of Ref. [2] obtain

$$\langle A_i^a(x^-, \vec{x}) A_i^a(x'^-, \vec{x}') \rangle = \frac{4(N_c^2 - 1)}{N_c |\vec{x} - \vec{x}'|^2}\left[1 - \left(|\vec{x} - \vec{x}'|^2 \Lambda_{\text{QCD}}^2\right)^{(N_c/16\pi)\chi(x^-, x'^-)|\vec{x} - \vec{x}'|^2}\right]. \tag{10}$$

This correlation function diverges like $(\vec{x}^2)^{\vec{x}^2}$ for large values of the separation. Of course, the bad behavior does not begin until the point $x \sim \Lambda_{\text{QCD}}^{-1}$, the point where we begin to mistrust our calculation anyhow. Unfortunately, because of this divergence, the Fourier transform of Eq. (10) does not exist for any value of $\vec{q}$, real or imaginary. So it is difficult to see how to define the gluon number density more than qualitatively using this expression.

The resolution of this problem lies in recognizing the importance of enforcing the observation that nucleons, when observed on a large enough length scale, should be color neutral. A consequence of the Gaussian averaging employed in the MV model is that the average color charge vanishes, $\langle \rho^a(x^-, \vec{x}) \rangle = 0$. However, we should also impose the (stronger) condition that

$$\int dx^- d^2\vec{x} \, \rho^a(x^-, \vec{x}) = 0 \tag{11}$$

for a nucleus-sized volume. If we integrate the charge density correlator (4) over all (x, x') and apply (11) we obtain a constraint on the transverse portion of the correlator $\mathcal{D}$:

$$\int d^2\vec{x} \, \mathcal{D}(\vec{x}) = 0. \tag{12}$$

Any correlation function which satisfies Eq. (12) is compatible with color neutral nucleons. Such a correlation function must contain an intrinsic scale, *i.e.* the minimum transverse length scale for which (12) becomes true. When we compute $L(\vec{x})$, this scale will be imparted to the logarithms appearing in Eq. (7). This stongly suggests that the resulting correlation function determined from Eq. (5) will be infrared finite. Indeed, this is the case [3].

To illustrate the features of our improved treatment, we turn to a specific model of a large nucleus introduced by Kovchegov [5]. In this model, we view the nucleus as containing A nucleons of radius a distributed within a sphere of radius R. Each "nucleon" consists of a $q\bar{q}$ pair. By explicitly averaging over the allowed positions of the quarks, antiquarks, and nucleons we may explicitly compute the function $\mathcal{D}(\vec{x} - \vec{x}')$. We find that there are two types of terms. The first is generated when the position of two quarks (or two antiquarks) overlap. It is proportional to $\delta^2(\vec{x} - \vec{x}')$, precisely the form for $\mathcal{D}$ employed in Ref. [2]. The second kind of term enters in with opposite sign and corresponds to the situation when a quark overlaps an antiquark. It is proportional to a smooth function of the separation. Since the focus of Ref. [5] was on very short distances, this term was neglected relative to the delta-function contribution. However, at somewhat longer distance scales, it is precisely this additional contribution which is required to satisfy Eq. (12).

We have compiled a series of plots (Figs. 3–6) to aid in the comparison of our results using Kovchegov's model to the results of Ref. [2]. In addition to the uniform distribution of quarks, antiquarks, and nucleons employed by Kovchegov, we have also performed the averaging using Gaussian distributions. In preparing these plots, we have adjusted the nucleon size parameter a and (for the MV result) Λ_{QCD} so that the corresponding correlation functions (5) match in the ultraviolet limit. This requires $a_G = 0.464 a_U$ and $\Lambda_{\text{QCD}} = 1.44 a_U^{-1}$ where a_U is the nucleon size parameter for the uniform quark distribution.

In Fig. 3 we have plotted the smooth part of the correlation function for the uniform and Gaussian cases, defined by writing $\mathcal{D}(\vec{x} - \vec{x}') \equiv \delta^2(\vec{x} - \vec{x}') - C(\vec{x} - \vec{x}')$. In both cases, the bulk of the contribution comes from separations less than $2a$.[1] Therefore, in this and subsequent plots we have defined the dimensionless distance $X \equiv |\vec{x} - \vec{x}'|/(2a)$.

Next, we present Fig. 4, which shows the trace of the correlation function (5) for all three cases. The Fourier transform of this quantity is proportional to the gluon number density. The Gaussian curve, not surprisingly, has a Gaussian tail for $X \geq 2a$, while the uniform curve vanishes identically in this region. On the other hand, the result from Ref. [2] runs off to $-\infty$ beyond $\Lambda_{\text{QCD}}^{-1}$.

We come to the gluon number density in Fig. 5. To define the MV curve, we have followed the suggestion made in Ref. [6], and simply cut off the $\vec{x}$ integration at $x = \Lambda_{\text{QCD}}^{-1}$. This is the source of the wiggles visible in Fig. 5. All three curves have the same overall shape, with a plateau at small values of $\vec{q}$ and a $1/\vec{q}^2$ fall off at large $\vec{q}$. The number of zero momentum gluons clearly depends on how we have cut things off: the Gaussian model, which allows quarks to be (albeit with small probability) a large distance from the center of their nucleons has the most long wavelength gluons, where as the MV curve, generated with a hard cutoff, has the fewest.

Finally, we display the effect of the non-Abelian terms on the gluon number

---

[1] In fact, $C(\vec{x} - \vec{x}')$ vanishes identically for separations greater than $2a$ when a uniform distribution is assumed.

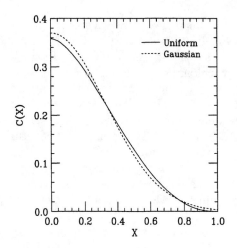

FIGURE 3. The smooth part of the two-point charge density correlation function in Kovchegov's model, defined by writing $\mathcal{D}(\vec{x}) \equiv \delta^2(\vec{x}) - C(\vec{x})$. The nucleon size parameters have been chosen so that the resulting gluon number densities match in the ultraviolet limit. The two curves correspond to uniform and Gaussian distributions of the quarks inside the nucleons.

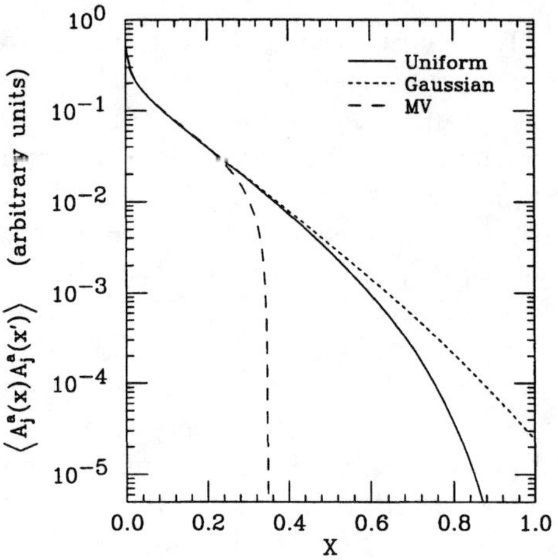

FIGURE 4. The trace of the two-point vector potential correlation function (5) in position space. The nucleon size parameters a and $\Lambda_{\rm QCD}$ have been chosen so that the resulting gluon number densities match for $X \to 0$. We have fixed the longitudinal coordinates at a place where $N_c \chi = 50 a^{-2}$. Plotted are the results of Ref. [2] (labelled "MV") as compared to Kovchegov's model [5].

density. This is most easily seen on a plot of $\vec{q}^{\,2}$ times the number density (Fig. 6).[2] From Eq. (5), it is clear that the magnitude of $N_c \chi$ governs the importance of these contributions. In Fig. 6 we have taken $N_c \chi$ to be 0, $50 a^{-2}$, and $100 a^{-2}$. What we see in this plot is a transfer of gluons from low values of $\vec{q}$ to higher ones. In fact, it is possible to show that the area under this plot is conserved [3], and that this shifing of gluons from one energy to another is the only effect of the non-Abelian contributions.

In conclusion, we see that imposing a color-neutrality condition on the nucleons eliminates the divergent infrared behavior of the two-point vector potential correlation function in the MV model. Because we have obtained a well-defined expression for the gluon number density, we are able to perform a quantitative investigation of the features of the MV model. For example, we have shown that the MV model predicts a gluon number density which is proportional to $1/q^+$ to all orders in the charge density. We are able to compute the gluon structure function in this framework, including its absolute normalization. The details of these and other related results may be found in Ref. [3].

---

[2] The fact that this curve approaches a constant for large $\vec{q}^{\,2}$ demonstrates the $1/\vec{q}^{\,2}$ behavior described in the text.

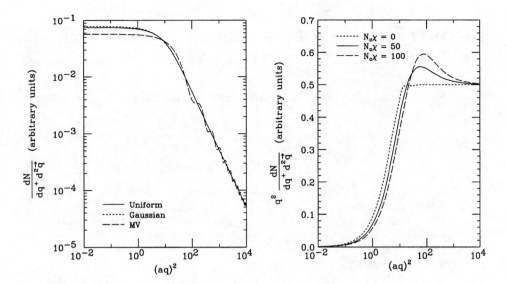

FIGURE 5. Plot of the gluon number density at fixed q^+. The nucleon size parameters a and $\Lambda_{\rm QCD}$ have been fixed so that these functions match for $\vec{q}^2 \to \infty$. We have chosen $N_c\chi = 50a^{-2}$. Plotted are the results of Ref. [2] (labelled "MV") as well as results using Kovchegov's model [5].

FIGURE 6. Plot of $\vec{q}^2$ times the gluon number density at fixed q^+ for various values of the color charge density within the uniform version of Kovchegov's model [5]. As χ increases, the importance of the non-Abelian contributions increases. Conversely, $\chi \to 0$ corresponds to the Abelian limit.

REFERENCES

1. L. McLerran and R. Venugopalan, Phys. Rev. **D49**, 2233 (1994); **D49**, 3352 (1994); **D50**, 2225 (1994); A. Ayala, J. Jalilian-Marian, and L. McLerran, Phys. Rev. **D52**, 2935 (1995).
2. J. Jalilian-Marian, A. Kovner, L. McLerran, and H. Weigert, Phys. Rev. **D55**, 5414 (1997).
3. C.S. Lam and G. Mahlon, "Color Neutrality and the Gluon Distribution in a Very Large Nucleus," hep-ph/9907281.
4. A.H. Mueller in *Frontiers in Particle Physics, Cargese 1994,* edited by M. Levy, J. Iliopoulos, R. Gastmans, and J.-M. Gerard, NATO Advanced Study Institute Series B, Physics, Vol. 350, (Plenum Press, 1995).
5. Yu. Kovchegov, Phys. Rev. **D54**, 5463 (1996).
6. L. McLerran and R. Venugopalan, Phys. Lett. **B424**, 15 (1998).

Quark- and Gluon-Condensate Contributions to Penguin Four-Fermi Operators

Mohammad R. Ahmady[1], Victor Elias[2]

Department of Applied Mathematics, University of Western Ontario
London, Ontario, Canada N6A 5B7

Abstract. The nonperturbative content of the QCD vacuum permits the occurrence of QCD-vaccum condensate contributions to penguin amplitudes. We calculate the dimension-4 $<m_q\bar{q}q>$ and $<\alpha_s G^2>$ contributions to the effective Wilson coefficients for penguin four-Fermi operators, and discuss how such contributions may contribute to nonleptonic B decays.

Nonleptonic B decays are very important to our understanding of standard model physics. In particular, decays of the form $B \to h_1 h_2$, where h_1 and h_2 are light hadrons π, K, η, η', ..., necessarily entail an operator product expansion framework with field theoretical coefficients $C_i(\mu)$:

$$<h_1 h_2 |H_{\text{eff}}|B> = \frac{G_F}{\sqrt{2}} \sum_i V_{\text{CKM}}^i C_i(\mu) <h_1 h_2|O_i(\mu)|B> \ . \tag{1}$$

The standard model's heavy degrees of freedom associated with the top quark and the $W^\pm$ have been integrated out of the effective Hamiltonian appearing in (1). The validity and utility of this description, as well as the effective truncation of higher dimensional operators from the expansion (1), must ultimately be tested empirically. For the case $B \to h_1 h_2$, it is standard to focus on the dimension-6 four-Fermi operators O_{1-6}, where operators $O_{1u,1c,2u,2c}$ are current-current operators and O_{3-6} are penguin operators as given below, with q being $\{s,d\}$, respectively for $b \to \{s,d\}$ transitions[$L \equiv 1 - \gamma_5$, $R \equiv 1 + \gamma_5$]:

$$O_{1u} = \bar{q}_\alpha \gamma^\mu L u_\alpha \bar{u}_\beta \gamma_\mu L b_\beta \ , \quad O_{2u} = \bar{q}_\alpha \gamma^\mu L u_\beta \bar{u}_\beta \gamma_\mu L b_\alpha \ ,$$
$$O_{1c} = \bar{q}_\alpha \gamma^\mu L c_\alpha \bar{c}_\beta \gamma_\mu L b_\beta \ , \quad O_{2c} = \bar{q}_\alpha \gamma^\mu L c_\beta \bar{c}_\beta \gamma_\mu L b_\alpha \ ,$$
$$O_3 = \bar{q}_\alpha \gamma^\mu L b_\alpha \sum_{q'} \bar{q}'_\beta \gamma_\mu L q'_\beta \ , \quad O_4 = \bar{q}_\alpha \gamma^\mu L b_\beta \sum_{q'} \bar{q}'_\beta \gamma_\mu L q'_\alpha \ ,$$

---

[1] Email: mahmady2@julian.uwo.ca
[2] Email:velias@julian.uwo.ca

$$O_5 = \bar{q}_\alpha \gamma^\mu L b_\alpha \sum_{q'} \bar{q}'_\beta \gamma_\mu R q'_\beta \; , \;\; O_6 = \bar{q}_\alpha \gamma^\mu L b_\beta \sum_{q'} \bar{q}'_\beta \gamma_\mu R q'_\alpha \; . \tag{2}$$

Methodologically, there exists a genuine issue as to how to partition radiative corrections between coefficients and the matrix elements in (1). We take the usual approach of allowing the matrix elements $< h_1 h_2 | O_i(\mu) | B >$ in (1) to be evaluated at tree order, with all higher order diagramatic perturbative corrections incorporated in Wilson coefficients $C_i(\mu)$.

We are particularly concerned with condensate contributions to penguin diagrams, in which an intermediate-state quark-antiquark pair emits an off-shell gluon that creates a final-state quark-antiquark pair. Such contributions occur by virtue of the nonperturbative content of the QCD vacuum, which necessarily entails non-vanishing condensate-dependent expressions for QCD-vacuum expectation values of normal-ordered products of fields, quantities which would ordinarily vanish if the vacuum were purely perturbative [1]. Thus the purely perturbative diagram (Fig. 1) must be augmented by diagrams sensitive to the nonperturbative vacuum expectation values of quark-antiquark pairs (Fig. 2) and multiple gluon fields (Fig. 3).

The purely perturbative Fig. 1 contributions to penguin operators are already known to next-to-leading (NLL) precision via naive dimensional regularization within an $\overline{\text{MS}}$ context [2]. We choose to incorporate such corrections entirely within the Wilson coefficients C_{3-6}, which are now denoted to be "effective" Wilson coefficients as follows:

$$C_3 \to C_3^{\text{eff}} = C_3 - \frac{1}{6}\Delta C \; , \tag{3}$$

$$C_4 \to C_4^{\text{eff}} = C_4 + \frac{1}{2}\Delta C \; , \tag{4}$$

$$C_5 \to C_5^{\text{eff}} = C_5 - \frac{1}{6}\Delta C \; , \tag{5}$$

$$C_6 \to C_6^{\text{eff}} = C_6 + \frac{1}{2}\Delta C \; , \tag{6}$$

with

$$\begin{aligned}\Delta C = \frac{\alpha_s}{4\pi} &\left\{ \left[\frac{4}{3} + \frac{2}{3} Ln(\frac{m_s^2}{\mu^2}) + \frac{2}{3} Ln(\frac{m_b^2}{\mu^2}) - \Delta F_1(\frac{k^2}{m_s^2}) - \Delta F_1(\frac{k^2}{m_b^2}) \right] C_3 \right. \\ &+ \sum_{q'=u,d,s,c,b} \left[\frac{2}{3} Ln(\frac{m_{q'}^2}{\mu^2}) - \Delta F_1(\frac{k^2}{m_{q'}^2}) \right] (C_4 + C_6) \\ &\left. - \sum_{q'=u,c} \frac{V_{q'b} V_{q'q}^*}{V_{tb} V_{tq}^*} \left[\frac{2}{3} + \frac{2}{3} Ln(\frac{m_{q'}^2}{\mu^2}) - \Delta F_1(\frac{k^2}{m_{q'}^2}) \right] C_1 \right\} \; , \end{aligned} \tag{7}$$

$$\Delta F_1(z) \equiv -4 \int_0^1 dx\, x(1-x) ln\left[1 - zx(1-x)\right] \; . \tag{8}$$

An important feature of such purely perturbative corrections to penguin operators is their dependence not only on the internal quark mass $m_{q'}$, but also on the value of k^2 characterizing the final-state quark-antiquark pair. This behaviour is particularly important for CP violation, since $\Delta F(k^2/m_{q'}^2)$ acquires an imaginary part when $k^2 > 4m_{q'}^2$. Such an imaginary part has already been suggested as the origin of the strong phase necessary for the generation of direct CP asymmetry [2].

The occurrence of k^2-dependent effective Wilson coefficients is also a feature of nonperturbative corrections arising from QCD vacuum condensates. Such condensate effects, which are already understood via their QCD sum-rule duality to the hadron spectrum, enter radiative corrections to penguin-mediated processes via Figs 2 and 3. To obtain the quark-condensate contribution to such processes (Fig 2), we proceed by replacing the usual perturbative internal quark propagator (S^{P}) in Fig. 1 with the full quark propagator $S(p)$:

$$S(p) = S^{\mathrm{P}}(p) + S^{\mathrm{NP}}(p). \tag{9}$$

The nonperturbative contribution to the quark propagator (S^{NP}) is just

$$iS^{\mathrm{NP}}(p) = \int d^4x e^{ip.x} <\Omega| : \Psi(x)\bar{\Psi}(0) : |\Omega>, \tag{10}$$

where $|\Omega>$ is the nonperturbative QCD vacuum [3]. The nonlocal vacuum expectation value (vev) in eq. (10) can be expanded in terms of local condensates. We note that the nonlocal vev of two quark fields does not contain a gluon condensate component [4]. The quark condensate projection of the nonperturbative quark propagator is taken from Ref. [3]:

$$iS^{\mathrm{NP}}(p) = (2\pi)^4(\slashed{p} + m_q)F(p), \tag{11}$$

where the Fourier transform of $F(p)$ is expressed in terms of Bessel function

$$\int d^4p e^{-ip.x} F(p) = -\frac{<\bar{q}q>}{6m_q^2} \frac{J_1(m_q\sqrt{x^2})}{\sqrt{x^2}}, \tag{12}$$

with the following mass-shell property:

$$(p^2 - m_q^2)F(p) = 0. \tag{13}$$

The dimension-3 quark condensate $<\bar{q}q>$ is defined to be the vacuum expectation value of the normal ordered local two-quark fields, i.e.

$$<\bar{q}q> = <\Omega| : \bar{\Psi}(0)\Psi(0) : |\Omega>. \tag{14}$$

Using the Feynman rule of the eq. (11), one can obtain the nonperturbative $<\bar{q}q>$ contribution to the effective Wilson coefficients in a straightforward manner. The relevant Feynman diagram is illustrated in Fig. 2 where the nonperturbative quark propagator S^{NP} is depicted by a disconnected line with two dots. Here we concentrate on the loop portion of Fig. 2 which differs from the purterbative case of Fig. 1. Aside from the color factor, the vector current correlation function can be written as

$$\Pi_{\mu\nu}^{<\bar{q}q>}(k) = 2\int d^4p \frac{Tr[(\slashed{p}-\slashed{k}+m)\gamma_\mu(\slashed{p}+m)\gamma_\nu]F(p)}{(p-k)^2 - m^2 + i\epsilon} , \qquad (15)$$

where the factor 2 in front is due to two possible insertions of the nonperturbative quark propagator in the fermion loop. By contracting $p^\mu p^\nu$ into $\Pi_{\mu\nu}^{<\bar{q}q>}$ and using the identity [3]

$$\int d^4p\, k.p F(p) = i \lim_{\xi\to 0} \frac{d}{d\xi} \int d^4p\, e^{-i\xi k.p} F(p)$$

$$= -i \frac{<\bar{q}q>}{6m_q} \lim_{\xi\to 0} \frac{d}{d\xi} \frac{J_1(m_q\xi\sqrt{k^2})}{m_q\xi\sqrt{k^2}} = 0 , \qquad (16)$$

where the second line is derived from eq. (12), one can show explicitly the transversality of the correlation function in eq. (15). Consequently, one finds upon contracting $g^{\mu\nu}$ into $\Pi_{\mu\nu}^{<\bar{q}q>}$ and imposing the on-shell constraint (13) that

$$\Pi^{<\bar{q}q>\mu}{}_\mu(k^2) = 24(2m_q^2 + k^2) \int d^4p \frac{F(p)}{k^2 - 2k.p + i\epsilon} - 24 \int d^4p F(p) . \qquad (17)$$

The integrals appearing in eq. (17) are evaluated as follows [3]:

$$\int d^4p F(p) = -\frac{<\bar{q}q>}{12m_q} ,$$

$$\int d^4p \frac{F(p)}{k^2 - 2k.p + i\epsilon} = \frac{i<\bar{q}q>}{12m_q^2\sqrt{k^2}} \int_0^\infty \frac{d\eta}{\eta} e^{i\eta k^2 - \epsilon\eta} J_1(2\eta m_q \sqrt{k^2})$$

$$= -\frac{<\bar{q}q>}{6m_q k^2} \left[1 + \sqrt{1 - \frac{4m_q^2}{k^2}}\right]^{-1} , \qquad (18)$$

where the final line is derived by utilizing a tabulated integral [5]. Using the transversality of $\Pi_{\mu\nu}^{<\bar{q}q>}$, the result for the RHS of (15) is easily obtained:

$$\Pi_{\mu\nu}^{<\bar{q}q>} = -\frac{<\bar{q}q>}{3m_q^3}\left[1 - \left(1 + \frac{2m_q^2}{k^2}\right)\sqrt{1 - \frac{4m_q^2}{k^2}}\right](g_{\mu\nu}k^2 - k_\mu k_\nu) . \qquad (19)$$

This result is, in fact the $<\bar{q}q>$-contribution to the vector-current correlation function first derived in ref. [6]. The corresponding gluon condensate contribution to the vector-current correlator, which may be derived by several different methods [3,6], is given by

$$\Pi_{\mu\nu}^{<G^2>} = -\frac{\alpha_s <G^2>}{48\pi m_q^4}(g_{\mu\nu}k^2 - k_\mu k_\nu)\frac{\left(\frac{m_q^2}{k^2}\right)^2}{\left(1 - \frac{4m_q^2}{k^2}\right)^2}$$

$$\times \left[\frac{48\left(\frac{m_q^2}{k^2}\right)^2\left(1 - \frac{2m_q^2}{k^2}\right)}{\sqrt{1 - \frac{4m_q^2}{k^2}}} Ln\frac{\sqrt{1 - \frac{4m_q^2}{k^2}} + 1}{\sqrt{1 - \frac{4m_q^2}{k^2}} - 1} - 4 + \frac{16m_q^2}{k^2} - 48\left(\frac{m_q^2}{k^2}\right)^2\right]. \quad (20)$$

where

$$z \equiv k^2/m_q^2 \quad (21)$$

and where [6,7]

$$X(z) \equiv \frac{1}{(1-4/z)}\left[\int_0^1 dx \, ln[1 - zx(1-x) - i|\epsilon|] + 2\right]$$

$$= \begin{bmatrix} \frac{1}{\sqrt{1+4/|z|}} ln\left[\frac{\sqrt{1+4/|z|}+1}{\sqrt{1+4/|z|}-1}\right], & z < 0 \\ \frac{2\sqrt{4/z-1}}{(1-4/z)} tan^{-1}\left[\frac{1}{\sqrt{4/z-1}}\right], & 0 < z < 4 \\ \frac{1}{\sqrt{1-4/z}}\left(ln\left[\frac{1+\sqrt{1-4/z}}{1-\sqrt{1-4/z}}\right] - i\pi\right), & z > 4 \end{bmatrix} \quad (22)$$

This contribution diverges sharply at $z = 4$, a divergence which is *not* removable by the operator realignment suggested in ref. [8].

We see that both $<\bar{q}q>$ and $<G^2>$ contributions to penguin operators occur via their known contributions to the vacuum polarization functions [eqs. (19) and (20)] within Figs 2 and 3. *in exactly the same way* that the purely perturbative vacuum polarization within Fig. 1 also enters penguin operators. This latter contribution is responsible for the vacuum-polarization Feynman loop integral ΔF_1, as defined by (8), which appears explicitly in the penguin-operator correction ΔC [eq. (7)]. Consequently, the aggregate effect of these lowest-dimensional condensate contributions to penguin operators is obtained by retaining equations (3-7), but with ΔF_1 redefined so as to include $<\bar{q}q>$ and $<G^2>$ contributions to the vacuum polarization:

$$\Delta F_1(z) = -4\int_0^1 dx\, x(1-x) Ln\left[1 - zx(1-x)\right]$$

$$+ \frac{8\pi^2 <m_q\bar{q}q>}{3m_q^4}\left[1 - \left(1 + \frac{2}{z}\right)\sqrt{1 - \frac{4}{z}}\right] \quad (23)$$

$$+ \frac{\pi <\alpha_s G^2>}{6m_q^4}\frac{1}{z^2(1-\frac{4}{z})^2}\left[\frac{48(1-\frac{2}{z})}{z^2\sqrt{1-\frac{4}{z}}} Ln\frac{\sqrt{1-\frac{4}{z}}+1}{\sqrt{1-\frac{4}{z}}-1} - 4 + \frac{16}{z} - \frac{48}{z^2}\right].$$

Eq. (23) is expressed in terms of the renormalization-group invariant condensates (to one-loop order in α_s) whose magnitude are known from QCD sum-rule applications. For example, if z is much larger than one, as would be the case for very light quarks [eq. (21)], the nonperturbative contributions to (23) are approximately given by

$$\Delta F_1^{\text{nonperturbative}} \approx \frac{16\pi^2 <m_q\bar{q}q>}{k^4} - \frac{2\pi <\alpha_s G^2>}{3k^4}. \quad (24)$$

Using phenomenological values $<m_q\bar{q}q> = -f_\pi^2 m_\pi^2/4$ [9] and $<\alpha_s G^2> = 0.045$ GeV^4 [10], these nonperturbative contributions are seen to be three orders of magnitude smaller than corresponding purely perturbative contributions to ΔF_1.

Indeed, for $B \to h_1 h_2$ decays, we may anticipate that k^2 is typically between $m_b^2/4$ and $m_b^2/2$ [11]. This range, however, suggests sensitivity to the $z = 4$ singularity mentioned above when the intermediate-state quark q' in (7) is identified with the charmed quark [recall that z in (23) can be identified with $k^2/m_{q'}^2$ in (7)]. One cannot, therefore, discount the possibility that the gluon-condensate contribution to the penguin amplitude may be comparable to the purely perturbative contribution.

This result has possible ramifications for investigations of CP asymmetry in hadronic B decays. The gluon condensate contribution to (7)[via (23)] develops an imaginary part as $k^2 \to 4m_c^2$ from above that would dominate the strong phase of the effective Wilson coefficients (3-6). Interestingly, the $<m_q\bar{q}q>$ contribution to (23), though small, has an imaginary part over the range $0 < z < 4$, suggestive of an additional source of CP violation stemming from light quark-antiquark-pair intermediate states. This latter absorptive contribution has been discussed in other contexts as a manifestation of the Goldstone theorem [12]. The singularity of the gluon-condensate contribution to penguin operator coefficients at $z = 4$, however, suggests a resonance interpretation.

Consider, for example, the dileptonic B decays $B \to X_s \ell^+ \ell^-$. Such decays acquire a continuum charmed quark loop contribution, analogous to the quark-loop in Fig. 1, in addition to a resonance contribution associated with the $c\bar{c}$ resonances $J/\psi, \psi', ...$ [13]. This resonance contribution can be modeled by the presence of a resonance propagator $1/(M^2 - k^2 - iM\Gamma)$ in the long distance amplitude, M and k being respectively the resonance-mass and momentum-transfer to the the lepton pair. The presence of the resonance propagator leads to a peak at $k^2 = M^2$ dominating the branching ratio for this decay mode. For the $B \to h_1 h_2$ case, the $z = 4$ gluon-condensate singularity of ΔF_1 within ΔC [eq. (7)] suggests the occurrence of a "charmonium" resonance at $k^2 = 4m_c^2$. However, the fact that this intermediate state couples to a *gluon*, as opposed to the electromagnetic-penguin photon appropriate for $B \to X_s \ell^+ \ell^-$, necessarily implies that this $c\bar{c}$ resonance be a weakly bound *colour octet* state. We note that the idea of colour-nonsinglet intermediate states within QCD is quite standard; such states are, of course, quarks and gluons, which are prohibited from manifesting themselves as physical final states by additional dynamics (confinement). What is new here is the idea of a (similarly confined) colour-nonsinglet bound state also contributing to a nonleptonic process via QCD's nonperturbative content, the QCD-vacuum condensate.

We reiterate, however, that the $q\bar{q}$ momentum-transfer k^2 is *not* an observable in the $B \to h_1 h_2$ decay process, in contrast to the observable dilepton momentum-transfer in $B \to X_s \ell^+ \ell^-$. It should be noted, though, that $4m_c^2$ is not an unreasonable choice for the value of k^2 characterizing $B \to h_1 h_2$, and that the gluon-condensate component of the decay amplitude may well play an important role both in enhancing such decays and in providing a mechanism for CP violation

Acknowledgement

We are grateful for support from the Natural Sciences and Engineering research Council of Canada.

REFERENCES

1. P. Pascual and R. Tarrach, QCD: *Renormalization for the Practitioner*, Lecture Notes in Physics Vol. 194, H. Araki, J. Ehlers, K. Hepp, R. Kippenhahn, H. A. Weidenmüller and J. Zittartz, eds. (Springer, Berlin, 1984) pp. 168-191.
2. G. Kramer, W. F. Palmer and H. Simma, Nucl. Phys. **B428**, 77 (1994); Z. Phys. **C66**, 429 (1995).
3. E. Bagan, M. R. Ahmady, V. Elias and T. G. Steele, Z. Phys. **C61** 157 (1994); Phys. Lett. **B305**, 151 (1993).
4. V. Elias, T. G. Steele and M. D. Scadron, Phys. Rev. **D38**, 1584 (1988).
5. I. S. Gradshteyn and I. M. Ryzhik, *Table of Integrals, Series and Products* (Academic Press, Orlando, 1980) p 712, Eq (6.623.3).
6. E. Bagan, J. I. Latorre and P. Pascual, Z. Phys. **C32**, 43 (1986).
7. A. S. Deakin et al, Eur. Phys. J. **C4**, 693 (1998).
8. M. Jamin and M. Münz, Z. Phys. **C60**, 569(1993).
9. M. Gell-Mann, R. J. Oakes and B. Renner, Phys. Rev. **175**, 2195 (1968).
10. M. A. Shifman, A. I. Vainshtein and V. I. Zakharov, Nucl. Phys. **B147**, 519 (1979).
11. H. Simma and D. Wyler, Phys. Lett. **B272**, 395 (1991); N. G. Deshpande and J. Trampetic, Phys. Rev. **D41**, 2926 (1990)
12. V. Elias, J. L. Murison, M. D. Scadron and T. G. Steele, Z. Phys. **C60**, 235 (1993); V. Elias and Mong Tong, Z. Phys. **C66**, 107 (1995); V. Elias and K. Sprague, Int. J. Theor. Phys. **37**, 2767 (1998).
13. M. R. Ahmady, Phys. Rev. **D53**, 2843 (1996).

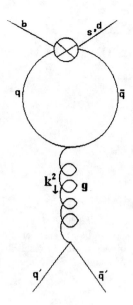

Figure 1

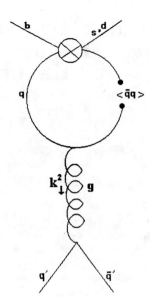

Figure 2

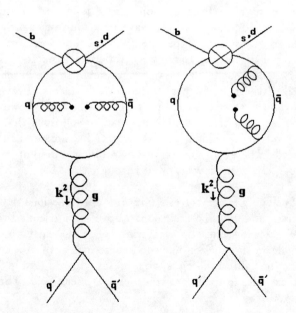

Figure 3

205

An investigation of nuclear collisions with a momentum-dependent Lattice Hamiltonian model

D. PERSRAM and C. GALE

Department of Physics, McGill University, 3600 University Street Montréal, Québec H3A 2T8, Canada

Abstract. We formulate a Lattice Hamiltonian approach for the modeling of intermediate energy heavy ion collisions. After verifying stationary ground state solutions, we implement this in a calculation of nuclear stopping power and compare our results with experimental data. Our findings support a relatively soft nuclear equation of state, with a momentum-dependent self-consistent mean field.

INTRODUCTION

For the past two decades the field of heavy ion collisions has seen many advances on both the experimental and theoretical fronts. At low and intermediate energies, much progress has come from the solution of transport theories such as the BUU [1–3] and QMD [4,5] models. Both have been quite successful in explaining many observables extracted from experimental studies at energies ranging from a few hundred MeV per nucleon up to a few GeV per nucleon. In this work, we concentrate on the first approach. Precise measurements in the lower energy regime ($\sim \epsilon_f$) have specified the need for precise numerical models. Previous BUU solutions have employed the so called "test particle" method [3]. However, it has been shown that at low energies this method may lead to solutions that can badly violate energy conservation [6]. This aspect is expected to worsen as the bombarding energy is decreased. The energy conservation problem inherent in the test particle method has largely been circumvented through the "Lattice Hamiltonian" [7] algorithm.

The process of colliding ions involves a mixture of interactions that the individual nucleons themselves undergo. For collisions above ϵ_f, nucleons can interact with each other via elastic and inelastic scattering. For the latter process to occur, the energy available in the nucleon-nucleon centre of mass frame must be at least $E = 2m_n + m_\pi$ since the lightest meson is the π meson. In addition to hard scattering, the nucleons also experience a self-consistent nuclear mean field. Thus, the nucleons will move on curved trajectories.

The mean field is a crucial ingredient in any transport calculation. Various *nuclear* mean fields have extensively been studied in the past [8–10]. In addition, it has previously been shown that different parameterizations of the nuclear mean field can yield similar results in measurements of transverse flow as can be seen in reference [8,9]. However transverse flow measurements do not exhaust the presently available experimental observations. In fact there are now available experimental data which can serve to distinguish between different parameterizations. This is the subject of this work.

NUCLEAR MATTER POTENTIALS

We begin our discussion by starting with nuclear matter considerations. The latter is defined as a net isospin zero infinite system of nucleons in which the total electric charge of the system is zero. Thus, isospin and Coulomb effects are not considered. We will use a semi-classical approximation which allows us to simultaneously specify the positions and momenta of all nucleons at all times. If we assume a smooth phase space distribution function $f(\vec{r},\vec{p})$, the total energy of the system reads:

$$\begin{aligned} E &= T + U \\ &= \int d^3r\, d^3p\, f(\vec{r},\vec{p}) \frac{p^2}{2m} \\ &\quad + \frac{1}{2} \int d^3\left(r,r',p,p'\right) f(\vec{r},\vec{p}) f(\vec{r},\vec{p}\,') v^{(2)}(\vec{r},\vec{r}\,',\vec{p},\vec{p}\,') + \frac{1}{3!} \int \cdots \end{aligned} \qquad (1)$$

In equation 1, $v^{(2)}(\cdots)$ represents a two-body nucleon-nucleon interaction. The potential energy term is written as a sum of n body interactions. Thus, our potential in general contains both two-body as well as many-body interactions. Next, we need to adopt a specific form for our n body interaction terms. A first simple choice is a momentum-independent contact interaction of strength a, and for the two-body direct term reads:

$$v^{(2)}(\vec{r},\vec{r}\,',\vec{p},\vec{p}\,') = a\delta(\vec{r}-\vec{r}\,'). \qquad (2)$$

If we lump the 3-body and higher interaction potentials into one term, we arrive at the "generalized Skyrme interaction" [11]. The corresponding potential energy *density* is shown below. The three parameters A, B and σ are left for us to choose, on the condition that we respect some constraints. We will discuss these in the next section.

$$W(\vec{r}) = \frac{A}{2} \frac{\rho^2(\vec{r})}{\rho_0} + \frac{B}{\sigma+1} \frac{\rho^{\sigma+1}(\vec{r})}{\rho_0} \qquad (3)$$

This generalized Skyrme interaction has been extensively studied in the context of heavy-ion collisions and has shown some success in describing a large amount of heavy-ion collision data at intermediate energies ($\sim$ 1GeV/A) with transport-type models [3,9,12,13]. However, there are other properties of nuclear matter that will manifest themselves during heavy-ion collisions which have yet to be unveiled. It is well known that the nuclear optical potential is strongly momentum dependent [14-16]. The simple phenomenological potential above does not contain any momentum-dependence. Thus, in order to obtain a more realistic description of nuclear matter one should include a term in the potential which includes some functional dependence on momentum. In this work, we use the Fourier transform of the finite range Yukawa potential. This momentum dependent interaction is then coupled with the zero range momentum-independent interaction and yields the following nuclear matter potential energy density shown below. This potential is known as the "MDYI" potential [17].

$$W(\vec{r}) = \frac{A}{2}\frac{\rho^2(\vec{r})}{\rho_0} + \frac{B}{\sigma+1}\frac{\rho^{\sigma+1}(\vec{r})}{\rho_0} + \frac{C\Lambda^2}{\rho_0}\iint d^3p\, d^3p'\, \frac{f(\vec{r},\vec{p})f(\vec{r},\vec{p}\,')}{\Lambda^2 + (\vec{p}-\vec{p}\,')^2} \qquad (4)$$

Skyrme and MDYI parameters

In the last section, two parameterizations of the nuclear mean field potential for nuclear matter were given. In both of those parameterizations, a number of "free" parameters were left unspecified. There are three for the Skyrme interaction and two additional ones (for a total of five) for the MDYI interaction. In order for these potentials to give a physical representation of nuclear matter, the value of each of the parameters must some how be connected to properties of nuclear matter. We use the experimentally observed properties of heavy ions to fix these parameters.

Let us first consider the momentum-independent Skyrme interaction. Two obvious conditions that should be satisfied are the binding energy per nucleon ($E_B = 16$ MeV) and the equilibrium density for nuclear matter($\rho_0 = 0.16$ fm^{-3}). Both of these are well established quantities [18]. A third condition that one can use to fix the three parameters is the determination of the nuclear matter compressibility (K). This quantity is directly related to the equation of state and gives a measure of the scalar elasticity of nuclear matter. For example, a soft or low value of the compressibility results in matter which is easily deformed by an external force while a stiff or high value provides matter which is relatively impervious to deformations. The value of the nuclear matter compressibility can be taken from the giant monopole resonance or breathing mode observed in heavy ions [19]. In addition, supernova calculations can provide additional constraints on this value [20]. The goal of this work is to attempt to deduce a value of the nuclear mean field compressibility from simulations of colliding heavy ions. We have chosen to use two values for the compressibility; a relatively soft EOS is provided with the choice of $K = 200$ MeV and a stiff EOS is provided with $K = 380$ MeV. These two values provide a reasonable bracket on this quantity as can be seen from supernova calculations as well as breathing mode calculations and observations [19-21].

The momentum-dependent MDYI interaction used in this work requires that we specify two more parameters. In equation 4, C represents the strength of the momentum-dependent term and Λ is representative of a range in momentum space. We now turn to nucleon-nucleus scattering experiments wherein one can extract information on the nuclear optical potential. This quantity is directly related to equation 4 as the latter contains information about the nucleon single particle potential $(U(\rho, \vec{p}))$ inside nuclear matter. By requiring $U(\rho = \rho_0, \vec{p} = 0) = -75$ MeV and $U(\rho = \rho_0, \vec{p}^2/2m = 300 \text{ MeV}) = 0$ we provide the two extra conditions necessary to fix all five parameters in the MDYI interaction potential. The agreement obtained with experimental measurements is very good from kinetic energies ranging from zero up to the GeV per nucleon regime [22]. With these parameterizations, we are now in a position to apply our nuclear transport model to the simulation of heavy ions collisions.

SIMULATION OF HEAVY IONS/COLLISIONS

In order to simulate colliding nuclei, there are still a few ingredients that we must add to the above nuclear matter approach. First, any stable nucleus contains a non-zero number of protons and is thus charged. We expect the Coulomb potential to play a role. Secondly, many heavy nuclei have a neutron number which can be as high as 1.5 times the atomic number: total isospin is non-zero. It is therefore also necessary to include an isospin potential into our formalism. We use an isospin potential that has previously been used in astrophysical considerations [23]. Now the total Hamiltonian of an A nucleon nucleus can be written down. The potential energy is discretized on a lattice(δx) in configuration space. We show below this Hamiltonian where α is a configuration space grid index.

$$H = \sum_{i=1}^{A} \frac{p_i^2}{2m} + (\delta x)^3 \sum_{\alpha} \left(W_\alpha + W_\alpha^{coul} + W_\alpha^{iso} \right) \quad (5)$$

In the above, W_α can either be the Skyrme or MDYI potential energy density.

As our Hamiltonian in equation 5 has no explicit time dependence, energy is conserved and we can now calculate the binding energy per nucleon for any size nucleus. We performed such a calculation for ~ 30 nuclei ranging from mass number $A:4 \rightarrow 260$ for both the Skyrme and MDYI potential energy densities. We obtain very good agreement with the Weizacker semi-empirical mass formula [18] over a large mass range. In passing, we note that the absence of an *explicit* surface potential yields about 1 MeV/A too large a binding energy for light nuclei. We do not expect this to significantly alter the results of this work.

The above analysis was done for a stationary nucleus. However, in heavy ion collision physics, one is concerned with the interaction of nuclei. Thus, we would like to know how the nucleons that make up two colliding nuclei evolve in time. We want to study the dynamical, *non-equilibrium* behaviour of colliding nuclei. In other words, we would like to have the equations of motions for all nucleons in this scenario. Obtaining this is a simple matter since we have the total Hamiltonian of the system. Thus, one has Hamilton's equations.

$$\dot{\vec{r}}_i = \nabla_{\vec{p}_i} H \qquad \dot{\vec{p}}_i = -\nabla_{\vec{r}_i} H \qquad (6)$$

The coupled set of nonlinear equations 6 together with equation 5 represent the *Lattice Hamiltonian* solution for colliding nuclei [7].

So far, we are able to calculate the trajectories of all nucleons in a time varying self-consistent mean field. However, as mentioned in the introduction one must also allow for elastic and inelastic nucleon-nucleon collisions as the total nucleon-nucleon cross section is in general non-zero. For the application of the model we have developed so far we will only be concerned with energies below the particle production threshold thus we need only consider the elastic nucleon-nucleon cross section. As we have included an isospin potential in our formulation, we will be using an elastic scattering cross section that is parameterized in terms of isospin and centre of mass energy [24]. For a detailed prescription of the scattering procedure used in this work the reader is referred to the reference by Bertsch and Das Gupta [3]. The details for solving Hamilton's equations in our model can be found in reference [25].

Now we are in a position to test the predictive power of the model. This is the subject of the following section.

NUCLEAR STOPPING

In order to give a qualitative picture of nuclear stopping it is instructive to consider two extreme examples of colliding nuclei. One can envisage that as two heavy ions approach each other on a collision course, there is a possibility for the two nuclei to coalesce. As two nuclei approach they are slightly slowed down by the Coulomb barrier. If the incident energy is just above that of the Coulomb barrier, the nuclei can merge into one large "nucleus". If the initial energy is sufficiently low such that there is no large buildup of density, repulsive mean field forces are at a minimum and a large compound nucleus will remain(assuming the Coulomb forces are not large enough to fission the nucleus). On the other hand, if one considers a high energy collision, there are regions where the matter density builds up rapidly and creates domains of large (positive) energy density. This in turn produces large pressure gradients which tend to expel the nucleons, thus breaking up the transient system. In the end we are left with many small remnants. At energies between these two extremes, experiments tell us that the final state can consist of a relatively large remnant with many smaller remnants in the final state. In order to quantify the stopping power and to compare with a specific experimental measurement, we will consider only the largest of these remnants. If we move to the lab frame, in the case of a single remnant, the latter will have a velocity equal to the velocity of the centre of mass of the two nuclei. At higher energies when there are more than one remnant present the velocity of the heaviest remnant will be smaller than in the previous case. This is due to the many small remnants which carry away a portion of the initial momentum. Thus, for a large final state remnant velocity, we have large stopping or close to complete absorption of the projectile. For a small final state remnant velocity, there is little stopping of the projectile as it partially rifles through the target. This description is exactly what is termed "nuclear stopping". Our goal here is to quantitatively investigate what role the nuclear matter compressibility as well as the momentum-dependence of the nuclear mean field play in determining the stopping power of nuclei.

Simulation results

Recent nuclear stopping results have been obtained at the NSCL at MSU using the K1200 cyclotron [26]. The longitudinal lab frame velocity of the heaviest remnant was determined for ^{40}Ar+^{108}Ag. The laboratory beam energies studied there ranged from $\sim 8 \rightarrow 115$ MeV/A. The experimental impact parameters were estimated from charged particle multiplicity and for the Ar+Ag system corresponds to $b \sim b_{max}/4$ [27]. In an attempt to bracket the experimental data we have performed Lattice Hamiltonian simulations for the Ar+Ag system at $b = b_{max}/3$ and $b = b_{max}/5$ for bombarding energies ranging from $\sim 20 \rightarrow 120$ MeV/A. The calculations were performed with both the Skyrme and MDYI nuclear mean field potentials as well

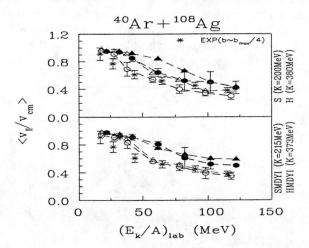

FIGURE 1. Fractional longitudinal laboratory frame velocity of the heaviest post-collision remnant as a function of laboratory bombarding energy. The experimental measurements [26] are shown by asterisks and the Lattice Hamiltonian calculations are shown with open and solid symbols. The circles are for a soft EOS and the triangles are for a stiff EOS. The filled(open) symbols are for an impact parameter of $b = b_{max}/5 (b = b_{max}/3)$. The top panel is with a Skyrme interaction and the bottom panel is with the MDYI interaction.

as with a stiff and soft EOS. For the Skyrme(MDYI) interaction, the compressibilities were 200(215)MeV for the soft EOS and 380(373)MeV for the stiff EOS. These parameterizations have been used before in a work by Zhang, Das Gupta and Gale [28].

Figure 1 displays the result for the longitudinal lab frame velocity of the heaviest remnant for both the measurement and calculation. Several conclusions can be taken from this. First, we note that the momentum-dependent mean field result is less sensitive to the value of the nuclear matter compressibility than is the momentum-independent mean field result. Indeed, the MDYI potential shows little sensitivity to the compressibility for this observable. This decrease in sensitivity is not too surprising once one takes into account the fact that the momentum-independent mean field is driven by the nuclear compressibility only while the momentum-dependent mean field contains an additional dependence on the momentum distribution. We find that the most convincing result for the momentum-independent mean field is obtained with a soft EOS. Both EOS's for the momentum-dependent mean field show nice agreement with the data for the larger value of the impact parameter.

Another observable considered was the mass of the heaviest remnant. These results are shown in figure 2. Note that from these results we can now separate the two momentum-dependent mean fields. As far as the momentum-dependent mean field is concerned (bottom panel), we find better agreement with a soft EOS. The data are only slightly underestimated in this case and the trend is reproduced. The momentum-independent mean field result on the other hand favours a stiff EOS. This is in contrast to the stopping result and indicates that one cannot satisfy both observables with the same momentum-independent mean field.

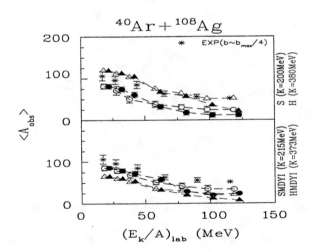

FIGURE 2. Observed mass of the heaviest remnant. All symbols and panels are as in figure 1.

CONCLUSION

We have implemented a Lattice Hamiltonian simulation for the case of two colliding heavy ions. Our solution incorporates both a momentum-independent as well as momentum-dependent nuclear mean field. The nuclear stopping results indicate that the momentum-dependent mean field is less sensitive to the nuclear matter compressibility than is the momentum-independent mean field. Furthermore, satisfactory agreement with both the stopping data and the observed large remnant mass can be achieved with the use of a momentum-dependent nuclear mean field. Our results support a relatively soft EOS of compressibility $K = 215$MeV.

REFERENCES

1. E. A. Uehling and G. E. Uhlenbeck, *Phys. Rev.* **43**, 552 (1933).
2. G. F. Bertsch, H. Kruse and S. Das Gupta, *Phys. Rev.* **C29**, 673 (1984).
3. G. F. Bertsch and S. Das Gupta, *Phys. Rep.* **160**, 189 (1988).
4. J. Aichelin and H. Stöcker, *Phys. Lett.* **B176**, 14 (1986).
5. J. Aichelin, *Phys. Rep.* **202**, 233 (1991).
6. C. Gale and S. Das Gupta, *Phys. Rev.* **C42**, 1577 (1990).
7. R. J. Lenk and V. R. Pandharipande, *Phys. Rev.* **C39**, 2242 (1989).
8. C. Gale, G.M. Welke, M. Prakash, S.J. Lee and S. Das Gupta, *Phys. Rev.* **C41**, 1416 (1990).
9. Qiubao Pan and Pawel Danielewicz, *Phys. Rev. Lett.* **70**, 2062 (1993).
10. G.F. Bertsch, W.G. Lynch and M.B. Tsang, *Phys. Lett.* **B189**, 384 (1987).
11. T.H.R. Skyrme, *Nucl. Phys.* **9**, 615 (1959).
12. G.D. Westfall et. al., *Phys. Rev. Lett.* **71**, 1986 (1993)
13. M.B. Tsang et. al., *Phys. Rev.* **C53**, 1959 (1996).
14. B. Friedman and V.R. Pandharipande, *Phys. Lett.* **B100**, 205 (1981).
15. J.P. Jeukeune, A. Lejeune and C. Mahaux, *Phys. Rep.* **25C**, 83 (1976).
16. R. Mafliet, *Prog. Part. Nucl. Phys* **21**, 207 (1988).
17. C. Gale et. al., *Phys. Rev.* **C41**, 1545 (1990).
18. See, for example, Samuel S.M. Wong *Introductory Nuclear Physics*, Prentice-Hall, Inc. Englewood Cliffs, New Jersey (1990).
19. D. H. Youngblood, H.L. Clark and Y.W. Lui, *Phys. Rev. Lett.* **82**, 691 (1999).
20. See, for example, H.A. Bethe, *Rev. Mod. Phys.* **62**, 801 (1990), and references therein.
21. M.M. Sharma, **nucl-th/9904036**, D. Vretenar et. al., **nucl-th/9612042**.
22. László P. Csernai, George Fai, Charles Gale and Eivind Osnes, *Phys. Rev.* **C46**, 736 (1992).
23. M. Prakash, T.L. Ainsworth and J.M. Lattimer, *Phys. Rev. Lett.* **61**, 2518 (1988).
24. J. Cugnon, D. L'Hôte and J. Vandermeulen, *NIM* **B111**, 215 (1995).
25. Declan Persram and Charles Gale, **nucl-th/9901019**.
26. E. Conlin et.al., *Phys. Rev.* **C57**, R1032 (1998).
27. R. Sun, private communication.
28. Jianming Zhang, Subal Das Gupta and Charles Gale, *Phys. Rev.* **C50**, 1617 (1994).

Lepton pair emission rates from a hot hadron gas

Ioulia Kvasnikova and Charles Gale

Physics Department, McGill University,
3600 University St., Montreal, QC, H3A 2T8, Canada

Abstract. We discuss the necessity and possible ways of dilepton pair emission rates calculations. We consider a relativistic kinetic theory approach and a finite temperature field theory approach. Qualitative comparison has been done.

INTRODUCTION

Experimental measurement of dilepton invariant mass spectra in heavy-ion collisions at intermediate and high bombarding energies have revealed a strong excess of pairs as compared to proton-induced reactions [1,2,3]. Being the ideal source of information about formation and observation of a quark-gluon plasma [4] those experimental data make it of our particular interest to investigate the possible sources of such behaviour. At CERN energies the collection of hadronic objects can successfully account for the measured dilepton spectrum in proton-induced reactions but fails to reproduce the experimental results in nucleus-nucleus collisions. Thus many theoretical efforts have concentrated on the role of medium effects on meson properties. In this talk we are concerned with interactions of light mesons in a hot baryon-free system and care must be taken of calculation of a corresponding purely hadronic signal.

For simplicity purposes we are dealing with the emission rate of dielectrons only but our discussion is completely general.

We shall make use of Vector Meson Dominance Model (VMD), which states that the hadronic electromagnetic current operator is given by the *current-field identity*

$$J_\mu = -\frac{e}{g_\rho}m_\rho^2 \rho_\mu - \frac{e}{g_\phi}m_\phi^2 \phi_\mu - \frac{e}{g_\omega}m_\omega^2 \omega_\mu.$$

We will restrict ourselves to ρ-meson propagation in a strongly interacting medium. But our study is completely general and ρ-meson can be replaced by ϕ- or ω-mesons.

To calculate medium properties in a hot meson gas we must consider the important and relevant resonances. Simple interaction Lagrangians, compatible with chiral symmetry and electromagnetic current conservation, are given by [7]

$$\mathcal{L}_{\rho PA} = G_{\rho PA}\, A_\mu (g^{\mu\nu} q\cdot p - q^\mu p^\nu)\rho_\nu P,$$

$$\mathcal{L}_{\rho PV} = G_{\rho PV}\, \epsilon_{\mu\nu\sigma\tau} k^\mu V^\nu q^\sigma \rho^\tau P,$$

$$\mathcal{L}_{\rho PP'} = G_{\rho PP'}\, P'(k\cdot q\, p_\mu - p\cdot q\, k_\mu)\rho^\mu P.$$

Here p^μ denote the four-momentum of the pseudoscalar, q^μ - ρ-meson and k^μ - axial-vector, vector or pseudoscalar mesons, respectively. At medium energies relevant for the hadronic gas phase, the light pseudoscalar Goldstone bosons $P = \pi, K$ are the most abundant species.

We fix the coupling constants using measured radiative decay widths together with VMD.

Now when we have defined our effective meson Lagrangians we can shift to possible treatment of ρ-meson propagation in hot meson matter.

RELATIVISTIC KINETIC THEORY APPROACH

The fundamental relativistic kinetic expression for the dilepton production rate (number of dilepton pairs per unit four-volume) for a process $a \to b + e^+e^-$ [6] is

$$\frac{dR_{a\to b+e^+e^-}}{dM^2} = \mathcal{N}\int \frac{d^3p_a}{2E_a(2\pi)^3}\frac{d^3p_b}{2E_b(2\pi)^3}\frac{d^3p_+}{2E_+(2\pi)^3}\frac{d^3p_-}{2E_-(2\pi)^3}$$

$$\times f_a\,(1+f_b)\mid \mathcal{M}\mid^2 (2\pi)^4$$

$$\times \delta^4(p_a - p_b - p_+ - p_-)\delta(M^2 - (p_+ + p_-)^2)$$

where M is the invariant mass of the e^+e^- pair, f's are Bose-Einstein mean occupation numbers and $\mathcal{N}$ is an overall degeneracy factor.

Using standard methods and spherical symmetry in momentum space this integral can be transformed to the form

$$\frac{dR_{a\to b+e^+e^-}}{dM^2} = \frac{\mathcal{N}m_a}{(2\pi)^2}\frac{d\Gamma_{a\to b+e^+e^-}}{dM^2}\int_{m_a}^\infty dE_a p_a f_a(E_a)$$

$$\times \int_{-1}^1 dx(1+f_b(E_b)),$$

with energy of b-particle E_b expressed as

$$E_b = \frac{E_a E_b^* + p_a p_b^* x}{m_a},$$

$$E_b^* = \frac{m_a^2 + m_b^2 - M^2}{2m_a}$$

and the differential decay width into the appropriate channel for $t = p_a - p_+$

$$\frac{d\Gamma_{a \to b + e^+ e^-}}{dM^2} = \int \frac{1}{(2\pi)^3} \frac{1}{32 m_a^3} \mid \mathcal{M} \mid^2 dt.$$

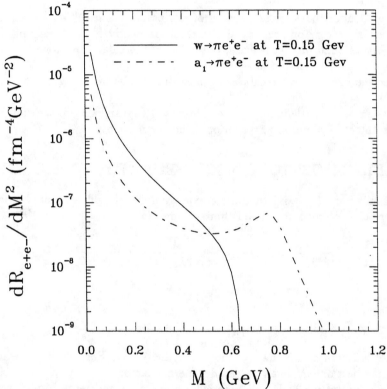

FIGURE 1. Dilepton pair production rate calculated using relativistic kinetic theory approach as a function of invariant mass for $\omega \to \pi + e^+ e^-$ (solid line) and $a_1 \to \pi + e^+ e^-$ (dashed-dotted line) at temperature 150 MeV.

In the rate expression, notice the term $(1 + f)$ which represents final state Bose-Einstein enhancement, an in-medium effect.

Considering processes $\omega \to \pi + e^+e^-$ and $a_1 \to \pi + e^+e^-$ and using our effective Lagrangians with coupling constants $g_{\omega\to\rho\pi}/g_{\rho\to\pi\pi} = 2.3\ GeV^{-1}$ and $g_{a_1\to\rho\pi}/g_{\rho\to\pi\pi} = 1.02\ GeV^{-1}$ we come to the dilepton pair production rates shown in Figure 1. For the graph $a_1 \to \pi + e^+e^-$ the peak around $M = 0.77\ GeV$ arises because of the ρ-meson form-factor contribution.

FINITE TEMPERATURE FIELD THEORY APPROACH

Our starting point is the model for ρ-meson in heat bath of mesons. Since the notion of temperature specifies a certain rest frame, Lorentz invariance is broken and the ρ-meson propagator splits into longitudinal and transverse modes and becomes a function of the energy q_0 and three-momentum $\vec{q}$ [7]

$$D_\rho^{\mu\nu}(q_0,\vec{q}) = \frac{P_L^{\mu\nu}}{M^2 - (m_\rho^{bare})^2 - \Sigma_\rho^L(q_0,\vec{q})}$$
$$+ \frac{P_T^{\mu\nu}}{M^2 - (m_\rho^{bare})^2 - \Sigma_\rho^T(q_0,\vec{q})} + \frac{q^\mu q^\nu}{(m_\rho^{bare})^2 M^2}$$

with the standard projection operators [5]

$$P_L^{\mu\nu} = \frac{q^\mu q^\nu}{M^2} - g^{\mu\nu} - P_T^{\mu\nu}$$

$$P_T^{\mu\nu} = \begin{cases} 0, & \mu=0 \text{ or } \nu=0 \\ \delta^{ij} - \frac{q^i q^j}{\vec{q}^2}, & i,j \in \{1,2,3\} \end{cases}$$

The longitudinal and transverse selfenergies are defined by the corresponding decomposition of the polarization tensor:

$$\Sigma_\rho^{\mu\nu}(q_0,\vec{q}) = \Sigma_\rho^L(q_0,\vec{q})P_L^{\mu\nu} + \Sigma_\rho^T(q_0,\vec{q})P_T^{\mu\nu}$$

that are calculated using Matsubara frequency formalism as

$$\Sigma_{\rho PR}^{L,T}(q_0,\vec{q}) = G_{\rho PR}^2 \int \frac{\vec{p}^2 d|\vec{p}|\,dx}{(2\pi)^2 2\omega_P(p)}$$

$$\times [f^P(\omega_P(p)) - f^{\rho P}(\omega_P(p)+q_0)]$$

$$\times F_{\rho PR}(q_{cm})^2 D_R(s)\ v_{\rho PR}^{L,T}(p,q)$$

where $v_{\rho PR}^{L,T}(p,q)$ is a projected vertex function given by above mentioned effective Lagrangian, $F_{\rho PR}(q_{cm})^2$ is a form-factor, $D_R(s)$ is a propagator of R-particle that can be an axial-vector, a vector or a pseudoscalar meson and f's are Bose-Einstein distribution functions.

The scalar form of individual spin-averaged self-energy contribution can be represented then as

$$\Sigma_{\rho PR}(M, \vec{q}) = \frac{1}{3}\,[\Sigma^L_{\rho PR}(M, \vec{q}) + 2\,\Sigma^T_{\rho PR}(M, \vec{q})]$$

Analogous expression is true for imaginary part of ρ-propagator

$$\mathrm{Im} D_\rho(M, \vec{q}; T) = \frac{1}{3}\,[\mathrm{Im} D^L_\rho(M, \vec{q}; T)$$

$$+\,2\,\mathrm{Im} D^T_\rho(M, \vec{q}; T)].$$

with

$$\mathrm{Im} D^{L,T}_\rho(M, \vec{q}; T) = \frac{\mathrm{Im}\Sigma^{L,T}_\rho(q_0, \vec{q})}{|\,M^2 - (m^{bare}_\rho)^2 - \Sigma^{L,T}_\rho(q_0, \vec{q})\,|^2}$$

So now when we have developed the formalism to calculate the imaginary part of ρ-propagator we are ready to go for dilepton production rate calculations.

The differential dilepton emission rate per unit four-volume and four-momentum in hot matter can be decomposed as [8]

$$\frac{dN_{l+l-}}{d^4 x\, d^4 q} = L_{\mu\nu}(q) H^{\mu\nu}(q).$$

If, for definiteness, we focus our attention on $e^+ e^-$ pairs then, since the electron-positron masses can be neglected compared to their three momenta, to lowest order in the electromagnetic coupling α, the lepton tensor $L_{\mu\nu}(q)$ takes the form

$$L_{\mu\nu}(q) = -\frac{\alpha^2}{3\pi^2 M^2}\Big(g_{\mu\nu} - \frac{q_\mu q_\nu}{M^2}\Big)$$

with the total pair four-momentum $q = p_+ + p_-$.

Within the VMD the hadronic tensor is directly related to the imaginary part of the retarded ρ-propagator in hot and dense matter

$$H^{\mu\nu}(q_0, \vec{q}; \mu_B, T) = -f^\rho(q_0; T)\,\frac{(m^{bare}_\rho)^4}{\pi g_{\rho\pi\pi}}\,\mathrm{Im} D^{\mu\nu}_\rho(M, \vec{q}; \mu_B, T)$$

The dilepton emission rate can then be written as

$$\frac{dN}{d^4 x\, d^4 q} = -\frac{\alpha^2 (m^{bare}_\rho)^4}{3\pi^3 g_{\rho\pi\pi}}\,\frac{f^\rho(q_0; T)}{M^2}$$

$$\times [\,\mathrm{Im} D^L_\rho(M, \vec{q}; T) + 2\,\mathrm{Im} D^T_\rho(M, \vec{q}; T)].$$

with the longitudinal and transverse spectral functions given above.

CONCLUSION

It's tempting to compare two different ways for the dilepton emission rate calculations. Finite temperature field theory technique has some apparent advantages over relativistic kinetic theory approach. Here are different reasons for that.

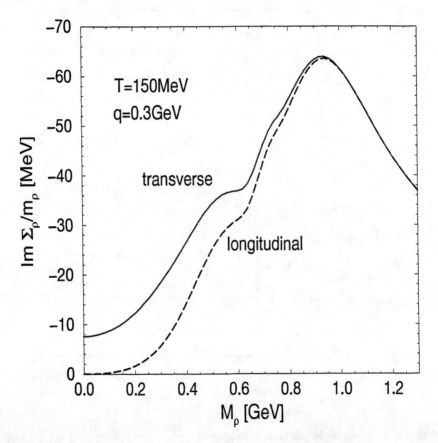

FIGURE 2. The longitudinal and transverse polarization contributions to the imaginary part of the ρ-meson self-energy [7].

First of all the existence of a preferred thermal reference frame will break Lorentz invariance. An advantage of theoretical approach like the one at hand is that the transverse and longitudinal parts of the ρ self-energy can be separately resolved. Even though at the present time there is no practical observable that is convincingly sensitive to the polarization, one should keep this difference in mind for future applications. From Figure 2 [7] we see that at finite three-momentum the longitudinal and transverse polarization contributions to the imaginary part of the ρ-meson self-energy summed over different channels, including baryonic ones, differ the most at low invariant masses and become undistinguishable at high masses.

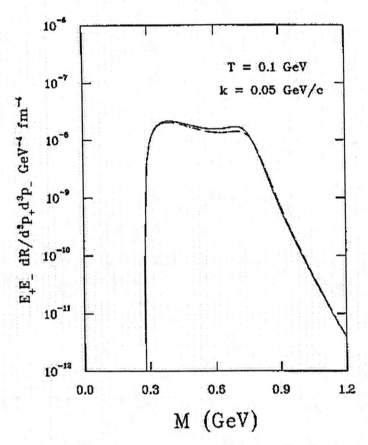

FIGURE 3. The rate of production for dilepton pairs versus their invariant mass M in a hot gas of pions. Solid curve correspond to the relativistic kinetic calculations and dashed and dashed-dotted - to the longitudinal and transverse contributions which are indistinguishable here [8].

Secondly summation of different processes in self-energy formalism is done in such a natural way that there is no need to take care of the interference effects as in our first approach. Longitudinal and transverse self-energy parts

$$\Sigma_\rho^L = \Sigma_{\rho\pi\pi} + \sum_\alpha \Sigma_{\rho\alpha}^L$$
$$\Sigma_\rho^T = \Sigma_{\rho\pi\pi} + \sum_\alpha \Sigma_{\rho\alpha}^T$$

where the summation is over the mesonic excitation channels α = $\pi\omega$, πh_1, πa_1, $\pi\pi'$, KK_1 etc. and $\Sigma_{\rho\pi\pi}$ also contains Bose-Einstein distribution functions.

I'd like to point out though that despite the obvious advantages of second approach, the results obtained within the finite temperature formalism does not differ much from kinetic theory calculations if we restrict ourselves to a purely pionic medium. Results shown in Figure 3 correspond to a single mesonic channel. For such a case the finite temperature self-energy put into expression for the kinetic theory thermal rate of the e^+e^--production give almost the same rates as in finite temperature formalism. The difference gets bigger as we go to higher total momentum. If we consider the medium rich in baryons as in the calculations represented in Figure 2 then the difference between the polarization states is more important. In this case, this is related to the nature of the resonances coupling to the ρ-meson.

REFERENCES

1. R.J.Porter et al., Phys.Rev.Lett.**79**, 1229 (1997).
2. G.Agakichiev et al., CERES collaboration, Phys.Rev.Lett.**75**, 1272 (1995), P.Wurm for the CERES collaboration, Nucl.Phys. **A590**, 103c (1995).
3. N.Masera for the HELIOS-3 collaboration, Nucl.Phys. **A590**, 93c (1995).
4. E.V. Shuryak, Phys.Lett. **78B**, 150 (1978).
5. J.I.Kapusta *Finite temperature Field Theory* (Cambridge University Press, Cambridge, England, 1989).
6. C.Gale and P.Lichard, Phys.Rev. **D49**, 3338 (1994).
7. R.Rapp and C.Gale, Phys.Rev. **C** in press.
8. C.Gale and J.Kapusta, Nucl.Phys. **B357**, 65 (1991).

The role of baryons in the production of dileptons during relativistic heavy ion collisions

A. K. Dutt-mazumder[1]*, C. Gale*, C. M. Ko† and V. Koch‡

*Physics Department, McGill University, Montreal, Quebec H3A 2T8, Canada
† Cyclotron Institute and Physics Department, Texas A&M University
College Station, TX 77843, USA
‡ Nuclear Science Division, Lawrence Berkeley National Laboratory
Berkeley, CA 94720, USA

Abstract. In the present work we investigate the baryonic contributions to the dilepton yield in high energy heavy ion collisions. The production rate of dielectron pairs from the Dalitz decay of low lying baryonic excitation are estimated. It has been observed that most dominant contribution comes from the decay of $N^*(1520)$ compared to other baryonic sources.

Production of dileptons in high energy nucleus nucleus collisions is one of the cardinal focus of contemporary heavy ion physics. This is largely due to the fact that electromagnetic radiation constitutes a penetrating spacetime probe of the production region: once produced it travels largely unscathed to the detectors. One can also show that the dilepton production rates are sensitive functions of the local density and temperature [1]. Of particular current interest is the excess yield of dielectrons in the low mass region ($M < 1$ GeV) observed in relativistic A-A collisions compared to what is seen p-A collisions. For instance, the CERES 95 data on S+Au and Pb + Au collisions, in the invariant mass region of $0.2 < M < 1.5$ GeV and $0.2 < M < 2.0$ GeV, show enhancement factors of $5.0 \pm 0.7 \pm 2.0$ and $3.5 \pm 0.4 \pm 0.9$ respectively [2]. A similar enhancement has also been reported by CERES in 96 where, for the Pb+Au system, the enhancement factor was found to be $2.6 \pm 0.5 \pm 0.6$ in the mass region $0.25 < M < 0.7$ [3]. A low mass enhancement has also been reported by the HELIOS/3 collaboration, for dimuons [4].

With respect to this data, a major theoretical challenge has therefore been to uncover the involved underlying physics which could result in such an enhancement. To this effect several attempts have been made in the recent past. A first generation of calculations involved meson secondary scattering. When the participating mesons have their vacuum properties, one found that the dilepton rates are indeed

---

[1] Speaker.

larger than those computed from radiative decay channels alone, but still not large enough to account for the data [5]. Two schools of thought currently claim success in reproducing the CERES data. The first idea attributes the dilepton excess to a considerable broadening of the rho meson in medium. This is achieved in part through coupling to heavier baryon resonances [6]. A second approach interprets the enhancement at low dilepton invariant mass as a signal of chiral symmetry restoration, through a dropping rho meson mass [7,8]. While a decisive experimental signature to distinguish between those two approaches still needs to be put forward, it is necessary to pursue detailed quantitative analyses in order to put any theoretical understanding of this phenomenon on a firm foundation.

In this light, we have decided to investigate the contribution of the principal low-lying baryon resonances to the lepton pair spectrum. We will start from precise electronuclear physics calculations. In doing this, we point out the importance of applying the knowledge gained in other nearby areas of subatomic physics to heavy ion collisions. Resonances considered here include $N^*(1520)$, $N^*(1720)$, $\Delta(1620)$, $\Delta(1700)$ and $\Delta(1905)$. Most important among these is the contribution of the $N^*(1520)$ [10]. For the sake of illustration and understanding the role of baryons we first focus exclusively on the Dalitz decay contribution of the $N^*(1520)$ to the dilepton spectrum.

To calculate the dilepton rate we consider the following hadronic Lagrangian

$$\mathcal{L}_{RN\gamma} = \frac{ieg_1}{2M}\bar{\psi}_R^\mu \Theta(z_1)_{\mu\nu}\gamma_\lambda \Gamma T_3 \psi_N F^{\nu\lambda}$$
$$- \frac{eg_2}{4M^2}\bar{\psi}_R^\alpha \Theta_{\alpha\mu}(z_2) T_3 \Gamma (\partial_\nu \psi_N) F^{\nu\mu} + h.c. \qquad (1)$$

where $\Theta_{\mu\nu}(z) = g_{\mu\nu} - 1/2(1+2z)\gamma_\mu\gamma_\nu$, Γ is either 1 or γ_5 depending upon the parity of the resonances and T is a $3/2 \to 1/2$ isospin transition operator or a 2×2 Pauli matrix, depending on the isospin of the resonance considered. Ψ_R^μ and ψ correspond to Rarita Schwinger and nucleon spinors, respectively, and $F^{\mu\nu}$ is the electromagnetic field tensor. The parameter z is associated with the off-shellness of spin 3/2 particles [11]. There is no unique way to determine these off-shell parameters which play a crucial role vis-a-vis the data corresponding to electromagnetic transitions of nucleon resonances into different multipolarities. It should be mentioned here that when the spin 3/2 particle is on-shell the results are independent of parameter z. Explicit calculation shows that the Rarita Schwinger projection operator

$$\Delta_{\mu\nu}(q) = (\slashed{q} + M_R)(-g_{\mu\nu} + \frac{2}{3}\frac{q_\mu q_\nu}{M_R^2} + \frac{1}{3}\gamma_\mu\gamma_\nu - \frac{1}{3}\frac{q_\mu\gamma_\nu - q_\nu\gamma_\mu}{M_R}) \qquad (2)$$

which appears in the squared matrix element for the decay of the 3/2 resonance contracted with z-dependent vertex terms γ_μ vanishes for on-shell particles. Or in other words only the first term of the $\Theta_{\mu\nu}(z)$ remains operative for vertex involving on-shell 3/2 resonances. The presence of the second term in the Lagrangian is important in order to keep the electric quadrupole (E1) and magnetic dipole (M2) transitions independent [11]. Quantitatively, however, the contributions from the second term to the dilepton yield are small by an order of magnitude compared to the first part.

Before going into the detailed discussions, we would like to point out the main differences between the present approach and what has been adopted in Ref [6] where the non-relativistic reduction of the following Lagrangian has been used to calculate the ρ spectral function :

$$\mathcal{L} = \frac{f_{RN\rho}}{m_\rho} \bar{\Psi}_R^\mu \gamma^\nu T F_{\mu\nu} \psi_N \qquad (3)$$

It is clear that here, unlike [6,10] and Eq. (3), we employ a direct coupling of photon with the resonances. This is to be noted that if we take off-shell parameter $z = -1/2$ and drop the second term, Eq.(1) can be cast into the form of Eq.(3) by invoking the Vector Meson Dominance (VMD) model. Nevertheless, the following points are to be noted in order to appreciate the difference of approach and method followed here. First, to deal with the vertex involving a spin 3/2 particle a systematic procedure has to be employed with regard to the off-shell corrections. This is built in the present approach. The importance of these off-shell parameters has been discussed at length in [11,12]. Secondly, we work within the framework of a fully relativistic formalism. It might be mentioned that the relativistic effects are found to be larger, as explained later, in the lower invariant mass region where baryonic contributions are more compared to that in the higher mass region.

Among the resonances considered here, we find, as mentioned before, that $N^*(1520)$ is the most dominant one. Therefore, for illustration, first we confine ourselves only to $N^*(1520)$. For the parameter set, we use the values from [12], where these are fitted to pion photoproduction multipoles. To be quantitative we take $g_1 = -1.839$, $g_2 = 0.018$, $z_1 = -0.092$ and $z_2 = -0.024$, while there are other parameter set also which gives rise to same χ^2 fit as far as the photo production data for different multipolarites are concerned, but differ to a large extent when we calculate the $N^*(1520) \to N\gamma$. The above quoted values of the parameter set gives the following decay widths : $N^*(1520) \to N\gamma = 0.78$ MeV and $N^*(1520) \to N\rho = 22.35$ MeV, while experimental values are 0.55±0.1 MeV and 24.0 ± 6.0 MeV respectively. It might be mentioned that when we use the non-relativistic version of Eq. (3), the width corresponding to radiative decay turns out to be 0.88 MeV and that for ρ is 24 MeV with coupling strengths $f_\rho = 5.5$ and $g_\rho = 6.0$, while a relativistic treatment with the same parameter set gives the width for $N\rho$ channel 30. MeV and for photon channel 1.13 MeV.

To estimate the rate for Dalitz decay of resonances we need to calculate $d\Gamma/dM^2$ from Eq. (1). Fig. 1 below depicts the results for free differential decay widths as a function of invariant mass (M) of the dilepton pairs. The dotted and solid curve correspond to the cases where the full Lagrangian i.e. Eq.(3) and its non-relativistic reductions have been used [6]. While the lower dashed-dotted curve depicts the result using Eq. (1). It might be recall from the preceeding paragraph that the photon branching ratios also show a similar trend for these three different cases and in consistent with the dielectron production rate. As far as the dilepton yield is concerned, the relativistic corrections, as evident from Fig. 1, are found to be important near the low invariant mass region while they tend to diminish at higher invariant masses. Physically this means as we move closer to the photon point, as expected, relativistic effects show up while the non-relativistic approximation, which, in effect, assumes the nucleon momenta to be small, becomes reasonable for higher invariant mass of the dilepton pair. Nevertheless, we note that the qualitative and quantitative trend and tendencies of the dilepton spectra are within reasonable limit, which might justify the use of non-relativistic reduction.

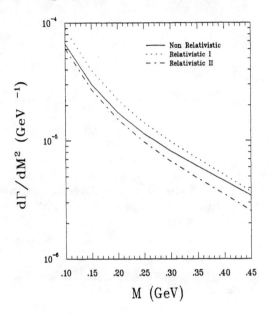

Fig. 1 : Solid and dotted line represent results corresponding to the non-relativistic and relativistic calculations using Eq.(3). The dashed-dotted line below shows the result corresponding to Eq (1.)

In order to investigate the effect of off-shell parameters we calculate $d\Gamma/dM^2$ both for $N^*(1520)$ on-shell and for the case when we integrate over the relevant spectral function of this resonant state. Fig. 2. shows the results corresponding to an on-shell $N^*(1520)$, and integrated over spectral function with and without the z-parameter dependent interactions. It is evident that here the off-shell effects arise not only from the off-peak mass distribution but also from the associated vertex containing the factor $\Theta_{\mu\nu}(z)$.

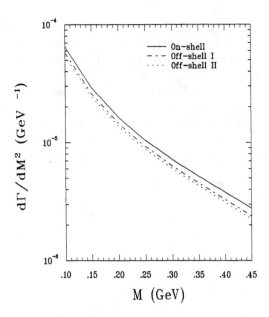

Fig 2. : The solid line shows results when $N^*(1520)$ is on-shell. The dashed-dotted and dotted curve depicts the results where the differential width has been integrated over $N^*(1520)$ spectral function with (Off-shell I) and with out off- shell interaction ($z = -1/2$) respectively (Off-shell II).

To obtain the rate of dilepton production for a given temperature and density from the Dalitz decay of resonances we employ the following equation. For any process $a \to b + e^+ e^-$ we have [1],

$$\frac{dR_{a \to b + e^+ e^-}}{dM^2} = \frac{g_a m_a}{(2\pi)^2} \frac{d\Gamma_{a \to b + e^+ e^-}}{dM^2} \int_{m_a}^{\infty} dE_a p_a f_a(E_a) \int_{-1}^{+1} dx (1 - f_b(E_b))] \quad (4)$$

where the detailed expressions for the kinematics are given in [1]. Here $f_i's$ represent relevant distribution functions in accordance with the spin of the particle. We should also mention here that, to incorporate proper off-shell dependence, we integrate the above equation over off-shell mass of the resonance weighted with the appropriate distribution function. We, however, treat the resonance decay width to be constant in this first exercise.

For completeness the individual contributions from the Dalitz decay of different baryonic resonances are shown in Fig. 3. In view of the fact that that the baryonic yield is much less compared to mesonic sources we take resort to the simpler interaction, i.e. Eq. (3), and take recourse to non-relativistic reduction to estimate individual contributions [10].

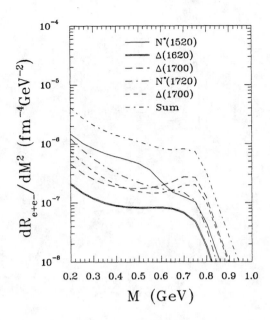

Fig. 3 The dilepton production rate for the process $R \to Ne^+e^-$ at T=170 MeV and zero pion chemical potential. From below to above curves correspond to $\Delta(1905)$, $\Delta(1620)$, $N^*(1720)$, $\Delta(1700)$, $N^*(1520)$ and their total contributions respectively.

To obtain the total dilepton yield for heavy ion collisions from static rate equations we adopt the simulation approach outlined in Ref. [13]. It is assumed that initially all particles are distributed uniformly within a cylinder of radius R_0 and longitudinal extent $2Z_l$. This somewhat simplified prescription could be improved by using a Gaussian distribution in the longitudinal direction. Nevertheless, the inclusion of formation time as discussed in [13] will effectively introduce the smearing of the particle density.

In Fig. 4, we present space-time integrated result where the contributions of various processes have been shown. Processes shown include Dalitz decay of π_0, η, a_1, ω and $N^*(1520)$, direct decay of ρ and ω meson into dielectron pair and $\pi - \pi$ annihilation. It is evident from Fig. 4. that $N^*(1520)$ contributes in a region where other mesonic sources give rise to much larger yield. Therefore, it seems reasonable to conclude that the Dalitz decay of baryonic resonances into e^+e^- is not of much significance. More detailed discussions and results of our calculations will be found in [14].

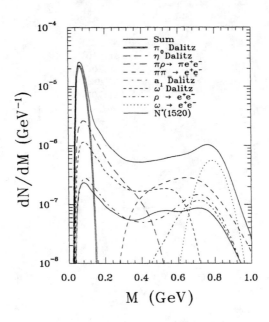

Fig 4.: Total dilepton yield as a function of invariant mass. Individual contributions from various processes are shown.

To conclude and summarize, in the current work, a model for dilepton production in HIC has been presented where the comparison has been made between the production rate coming from, mesonic sources, Dalitz decay of baryonic resonances and pion-pion annihilation. We find that the region where baryonic contributions are larger is mostly dominated by the mesonic sources as evident from Fig. 4. Still further investigations should be made in order to understand their role more precisely both at the tree level and higher order interactions.

ACKNOWLEDGMENT

CG wishes to acknowledge the hospitality of the Theory Institute of the University of Giessen, where this work began. The work of CG and AKDM was supported by the Natural Sciences and Engineering Research Council of Canada and the Fonds FCAR of the Québec Government. The work of CMK was supported in part by the National Science Foundation under Grant No. PHY-9509266 and PHY-9870038, the Welch Foundation under Grant No. A-1358, the Texas Advanced Research Program, and the Alexander Humboldt Foundation. VK was supported by the Director, Office of Energy Research, division of Nuclear Physics of the Office of High Energy and Nuclear Physics of the U.S. Department of energy under Contract No. DE-AC03-76SF00098.

REFERENCES

1. Charles Gale and Peter Lichard, Phys. Rev. D **49** (1994) 3338.
2. G. Agakichiev et al., Phys. Rev. Lett. **75**, 1272 (1995); Nucl. Phys **A610**, 317 (1996); A. Drees, Nucl. Phys. **A630**, 449c (1998), and references therein.
3. B. Lenkeit, in *Proceedings of INPC 98*, Paris Aug. 24-28, 1998, Nucl. Phys. A to be published.
4. M. Masera et al., Nucl. Phys. **A590** (1995) 93c.
5. W. Cassing, W. Ehehalt, and C. M. Ko, Phys. Lett. **B363**, 35 (1995); D. K. Srivastava, B. Sinha, and C. Gale, Phys. Rev. C **53**, R567 (1996); C. M. Hung and E. V. Shuryak, Phys. Rev. C **56**, 453 (1997); V. Koch and C. Song, Phys. Rev. C **54**, 1903 (1996); J. Murray, W. Bauer, and K. Haglin, Phys.Rev. C **57**, 882 (1998).
6. R. Rapp, G. Chanfray, and J. Wambach, Phys. Rev. Lett. 76 (1996) 368; Nucl. Phys. A 617 (1997) 472.
7. G. E. Brown and M. Rho, Phys. Rev. Lett. 66 (1991) 2720.
8. G. Q. Li, C. M. Ko, and G. E. Brown, Phys. Rev. Lett. 75 (1995) 4007; Nucl. Phys. A 606 (1996) 568; A 611 (1996) 539.
9. P. Huovinen and M. Prakash, Phys. Lett B 450(1999)15
10. W. Peters, M. Post, H. Lenske, S. Leupold, and U. Mosel, Nucl. Phys. A 632 (1998) 109.
11. M. Benmerrouche and Nimai C. Mukhopadhyay, Phys. Rev. D 51(1995)3237
12. T. Feuster and U. Mosel, Nucl. Phys. A 612 (1997)375
13. V. Koch and C. Song, Phys. Rev. C 54(1996)1903
14. A. K. Dutt-mazumder, C. Gale, C.M Ko, and V. Koch, in preparation.

MRST `99: Participants

Ahmady, Mohammad	Western	mahmady2@julian.uwo.ca
Armitage, John	Carleton	armitage@physics.carleton.ca
Bassi, Zorawar	Cornell	zorawar@mail.lns.cornell.edu
Batkin, I.	Carleton	batkin@physics.carleton.ca
Bauer, Christian	Toronto	bauer@physics.utoronto.ca
Bensalem, Wafia	Montreal	wafia@lps.umontreal.ca
Boutin, Jean-Guy	Carleton	boutin@physics.carleton.ca
Buchel, Alex	Cornell	buchel@mail.lns.cornell.edu
Carnegie, Robert	Carleton	carnegie@physics.carleton.ca
Catterall, Simon	Syracuse	smc@physics.syr.edu
Clarke, Robert	Carleton	clarke@physics.carleton.ca
Cline, James	McGill	jcline@hep.physics.mcgill.ca
Copley, Leslie	Carleton	lcopley@physics.carleton.ca
Couture, Gilles	UQAM	couture@mercure.phy.uqam.ca
Datta, Alakabha	Toronto	datta@medb.physics.utoronto.ca
Doncheski, Michael	Pennsylvania St.	mad10@psu.edu
Dremin, Igor	Lebedev P. I.	dremin@td.lpi.ac.ru
Dutt-Mazumder, A.K.	McGill	abhee@hep.physics.mcgill.ca
Fariborz, Amir	Syracuse	amir@suhep.phy.syr.edu
Fleming, Sean	Toronto	fleming@physics.utoronto.ca
Frank, Mariana	Concordia	mfrank@vax2.concordia.ca
Gerganov, Bogomil	Cornell	beg@mail.lns.cornell.edu
Godfrey, Steve	Carleton	godfrey@physics.carleton.ca
Gregory, Eric	Syracuse	gregory@phy.syr.edu
Groote, Stefan	Cornell	groote@mail.lns.cornell.edu
Hagen, Carl R	Rochester	hagen@urhep.pas.rochester.edu
Hamzaoui, Cherif	UQAM	hamzaoui.cherif@uqam.ca
Ivanovic, Igor	Carleton	igor@physics.carleton.ca
Kalyniak, Pat	Carleton	kalyniak@physics.carleton.ca
Kamal, Basim	Carleton	bkamal@physics.carleton.ca

MRST '99: Participants (continued)

Khanna, Faqir	Alberta	khanna@phys.ualberta.ca
König, Heinz	Concordia	konig@mercure.phy.uqam.ca
Kvasnikova, Ioulia	McGill	ioulia@hep.physics.mcgill.ca
Lalonde, Christiane	UQAM	lalonde@mercure.phy.uqam.ca
Lam, Harry	McGill	lam@physics.mcgill.ca
Lee, C.-W Herbert	Rochester	cwhlee@pas.rochester.edu
London, David	Montreal	london@lps.umontreal.ca
Macesanu, Cosmin	Rochester	mcos@pas.rochester.edu
Mahlon, Gregory	McGill	mahlon@physics.mcgill.ca
McGuire, Scott	Syracuse	svmcguir@syr.edu
Milek, Marko	McGill	milek@physics.mcgill.ca
Mouline, Saad	UQAM	saad@mercure.phy.uqam.ca
Newhouse, Jim	U. St. Thomas	jnewhouse@atcorp.com
Oakham, Gerald	CRPP/Carleton	oakham@crpp.carleton.ca
O'Donnell, Patrick J.	Toronto	odonnell@physics.utoronto.ca
Patel, Popat	McGill	patel@hep.physics.mcgill.ca
Persram, Declan	McGill	declan@hep.physics.mcgill.ca
Pospelov, Maxim	Minnesota	pospelov@hep.umn.edu
Raymond, Alexandre	UQAM	raymond@mercure.phy.uqam.ca
Resnick, Laser	Carleton	resnick@physics.carleton.ca
Romo, William J.	Carleton	romo@physics.carleton.ca
Sundaresan, M.K.	Carleton	sundaresan@physics.carleton.ca
Szalapski, Robert	Rochester	robs@pas.rochester.edu
Toharia, Manuel	UQAM	toharia@mercure.phy.uqam.ca
Torma, Tibor	Toronto	kakukk@physics.utoronto.ca
Waller, David	Carleton	dwaller@physics.carleton.ca
Watson, Peter	Carleton	watson@physics.carleton.ca

MRST '99: High Energy Physics at the Millenium

Registration & Welcome - Monday 9:00-10:30

I - Monday 10:30-11:30; New Physics Searches
Chair: Simon Catterall

Basim Kamal (Carleton U.)	Searching for a W' at the NLC using single photons
Michael Doncheski (Penn. State U.)	Closing the low-mass axigluon window

II - Monday 11:30-12:30; Physics of Hadrons
Chair: David London

Amir Fariborz (Syracuse U.)	Lowest-lying scalar mesons and a possible probe of their quark substructure
Maxim Pospelov (U of Minnesota)	Theta term in the QCD sum rule approach and the electric dipole moments of hadrons

Lunch 12:30-14:00

III - Monday 14:00-16:00; SUSY & GUT's
Chair: Mike Doncheski

Mariana Frank (Concordia U.)	Spectrum of gauge mediated breaking of SO(10)
Alex Buchel (Cornell U.)	Ultrastrong coupling in N=2 supersymmetric gauge theories
Cosmin Macesanu (U. of Rochester)	Charged lepton mixing and R-parity violating SUSY
Manuel Toharia (UQAM)	Higgs-mediated FCNC in supersymmetric models with large tan(beta)

Coffee 16:00-16:30

IV - Monday 16:30-18:00; Field Theory and Gravity I
Chair: Mariana Frank

Jim Cline (McGill U.)	Inflation with extra dimensions
C.-W. Herbert Lee (U. of Rochester)	Large-N Yang-Mills theory as a classical mechanics
Eric Gregory (Syracuse U.)	Gauge field correlators in 4-D simplical quantum gravity

V - Tuesday 9:00-10:30; Field Theory & Gravity II
Chair: Jim Cline

Bogomil Gerganov (Cornell U.)	The multi-cosine model -- algebraic structures & integrable points on the marginal manifold
Harry Lam (McGill U.)	Coherent state in high energy physics
Scott McGuire (Syracuse U.)	Ising model on fluctuating disk geometries

Coffee 10:30-11:00

VI - Tuesday 11:00-12:15; Neutrino & Electroweak Physics
Chair: Harry Lam

M.K. Sundaresan (Carleton U.)	Coherent conversion of neutrino flavour by collisions with relic neutrino gas
Stefan Groote (Cornell U.)	Local and global duality and the determination of alpha(M_Z)

Lunch 12:15-14:00

VII - Tuesday 14:00-15:30; B Physics and Jets
Chair: Patrick O'Donnell

David London (U. of Montreal)	Looking for new physics in B_d - B_dbar mixing
Wafia Bensalem (U. of Montreal)	Standard Model T-odd asymmetries in B decays
Alakabha Datta (U. of Toronto)	Measuring the CKM angle beta in B -> $D^* D^* K_s$

Coffee 15:30-16:00

VIII - Tuesday 16:00-17:30; (continued)
Chair: Peter Watson

Christian Bauer (U of Toronto)	Infrared effects in B-decays
Patrick J. O'Donnell (U of Toronto)	Isospin predictions for the Λ_b decays
Igor Dremin (Lebedev Physical Inst.)	QCD on multiparticle production

Reception and Banquet - Cartier Place Towers - 18:30

IX - Wednesday 9:00-10:30; Collider Physics
Chair: Carl Hagen

Marko Milek (McGill U.)	Background suppression using various selectors
Robert Szalapski (U. of Rochester)	Sum rules for W-boson pair production at one-loop
Tibor Torma (U. of Toronto)	New physics in top polarization at the Tevatron

Coffee 10:30-11:00

X - Wednesday 11:00-13:00; Nuclei & Nonperturbative Physics
Chair: Faqir Khanna

Gegory Mahlon (McGill U.)	Gluons in a color-neutral nucleus
Mohammad Ahmady (U. of Western O.)	Nonperturbative contributions to the effective Wilson coefficients of the four-Fermi operators
Declan Persram (McGill U.)	An investigation of nuclear collisions with a momentum-dependent lattice Hamiltonian model
Ioulia Kvasnikova (McGill U.)	Lepton pair emission rates from hot gas of mesons
Abhee Dutt-Mazumder (McGill U.)	Dilepton production in relativistic heavy ion collisions

End of Conference-enjoy Ottawa!

AUTHOR INDEX

A

Ahmady, M. R., 198

B

Batkin, I. S., 110
Bauer, C. W., 154
Bensalem, W., 140
Bisset, M., 49
Browder, T. E., 146
Buchel, A., 40

C

Catterall, S. M., 80, 103
Cline, J. M., 64

D

Datta, A., 146, 161
Doncheski, M. A., 9
Dremin, I. M., 169
Dutt-Mazumder, A. K., 222

E

Elias, V., 198

F

Fariborz, A. H., 17
Frank, M., 33

G

Gale, C., 206, 214, 222
Gerganov, B., 86
Godfrey, S., 1
Gregory, E. B., 80
Groote, S., 124

H

Hamidian, H., 33
Hamzaoui, C., 56

K

Kalyniak, P., 1
Kamal, B., 1
Ko, C. M., 222
Koch, V., 222
Kong, O. C. W., 49
Kvasnikova, I., 214

L

Lam, C. S., 95
Lee, C.-W. H., 72
Leike, A., 1
Lipkin, H. J., 161
London, D., 132

M

Macesanu, C., 49
Mahlon, G., 190
McGuire, S. V., 103
Milek, M., 174

O

O'Donnell, P. J., 146, 161
Orr, L. H., 49

P

Pakvasa, S., 146
Patel, P. M., 174
Persram, D., 206
Pospelov, M., 25, 56
Puolamäki, K., 33

R

Rajeev, S. G., 72
Raymond, A., 56
Ritz, A., 25

S

Sundaresan, M. K., 110

T

Toharia, M., 56
Torma, T., 182